Construction Trades

INFORMATION PROCESSING SKILLS:

READING

Thomas G. Sticht
Barbara A. McDonald

GOALS
Glencoe Occupational Adult Learning Series

New York, New York   Columbus, Ohio   Mission Hills, California   Peoria, Illinois

This program was prepared with the assistance of
Chestnut Hill Enterprises, Inc.

Construction Trades Information Processing Skills: Reading

Imprint 1996

Send all inquiries to:

Glencoe/McGraw-Hill
936 Eastwind Drive
Westerville, OH 43081

ISBN 0-07-061534-9

3 4 5 6 7 8 9 10 11 12 POH 03 02 01 00 99 98 97 96

Table of Contents

Preface

People used to think that when they got out of school, they could stop learning. Today, that is no longer true!

People who want careers in well-paying jobs have to use their knowledge and skills to learn every day. They have to keep up with rapid changes in technology. They must meet new demands for goods and services from customers. They have to compete for good jobs with workers from around the world.

The books in this program will help you learn how to learn. The *Construction Trades Knowledge Base* will give you the background you need to learn about the construction trades. The *Reading and Mathematics Information Processing Skills* books will teach you how to use your skills to learn new information.

When you complete these three books, you will be ready for more training in the construction trades. Then when you start your career, you will be able to learn new knowledge and skills. This way, you will always be able to keep up with changes in jobs. You will also be ready to move ahead to jobs of greater responsibility.

We wish you the very best of success in your chosen career field!

PART 1 INTRODUCTION

The ability to read is one of the most important, most useful skills you can have. Reading is becoming more important every day. The reason is simple: No matter what kind of work you do, you *must* be able to read to do your job well.

❑ READING ON THE JOB

Today's construction world is an information world. You may work at a construction site, for a subcontractor, or in a factory. What you read on the job depends on the kind of work you do. You may read many letters and memos. Perhaps you will read charts and graphs. You may need to read reports and manuals. Chances are, you will need to read many different kinds of materials. Whatever you do, you will need good reading skills.

SELF-CHECK 1-1

Directions: *If you now work, take a separate sheet of paper and list the kinds of materials that you read on your job.*

❑ THE IMPORTANCE OF GOOD READING SKILLS

Good reading skills will make your work easier. When you can apply good reading skills, you will be able to find answers to on-the-job questions. You will be able to solve on-the-job problems.

For example, when you learn to use parts of books, such as the table of contents, you will be able to find information in books very quickly. You will save time, but more important, you will be sure to find the right answers.

SELF-CHECK 1-2

Directions: *In the space below, write three ways that good reading skills can help you in your job.*

1. ______________________________

2. ______________________________

3. ______________________________

SELF-CHECK 1-3

Directions: *In addition to the table of contents, can you name other parts of a book that might help you find something? Write your answers in the space below.*

____

____

This book and the *Construction Trades Knowledge Base* will help make your reading both easier and more effective by helping you to understand your reading goals. Let's see what this means.

❑ UNDERSTANDING YOUR GOALS

At times, you will read to *learn* something. Your goal will be to understand as many details of what you are reading as you can in order to master the information. At other times, you will read simply to *do* something—for example, to find an answer to a question. As you will see, reading to learn and reading to do are two very different goals, but both require reading skill.

READING TO LEARN

Assume that you must read a manual that describes how to use a new piece of equipment. The equipment can be dangerous if not used properly.

Of course, you want to learn as much about this equipment (especially its safety features) as you can. You want to *recall* the information you learn. That means that you want to be able to *use* the information in the future. You want to be able to apply what you read when you operate the equipment.

The information is important. Your purpose, your goal in reading, is to *master* the information. You want to keep the information in your permanent memory. Your goal is *reading to learn.*

READING TO DO

But you do not always read to master information. Sometimes, you will read simply to find an answer to a question or a problem. Once you find that answer, it really may not be important to *keep* the information in your memory at all.

For example, suppose you need to find the manufacturer's serial number for that same piece of equipment. You need to (1) find the number, (2) write it accurately on a form, and (3) give the filled-out form to your supervisor. Do you need to *memorize* the number? No. Do you need to *understand* it? No.

You simply need to *find* the number. It's somewhere in the 300-page manual that comes with the equipment. But you don't want to read every page of the manual just to find the number.

Of course, you *can* find the number quickly and easily. How? By using "finding tools" such as the table of contents in the manual. Your goal here is *reading to do.*

SELF-CHECK 1-4

Directions: *You have been asked to do the following tasks. In the space provided, put a check mark to indicate whether they are reading-to-learn or reading-to-do activities.*

Task	Reading to Learn	Reading to Do
1. Read a shipping chart to find the cost of mailing a three-pound package.	____	____
2. Read the instruction manual to learn how to use a radial-arm saw.	____	____
3. Read a chapter in a textbook for a homework assignment.	____	____
4. Read a parts manual to find the number and cost of a part.	____	____
5. Read a telephone book to find a friend's number.	____	____
6. Read a manual to learn how to use a VCR.	____	____
7. Read a book about the Vietnam War.	____	____
8. Read the newspaper to find the score of last night's basketball game.	____	____
9. Read the newspaper to find out about an earthquake in California.	____	____
10. Read your *Construction Trades Knowledge Base.*	____	____

As you use this book together with your *Construction Trades Knowledge Base*, you will learn more about reading-to-learn and reading-to-do skills. First, let's look at the system your mind uses to process information. As you will see later, this will help you to use reading-to-learn and reading-to-do skills.

❑ INFORMATION PROCESSING

How do we learn? How does new information enter the memory? How does our brain remember? How does our mind store information and then find it again? In other words, how does the brain *process* information?

Understanding how we learn and how our brains process information will help you to improve your reading skills. To understand this process, let's begin by taking a look at the brain and its "inventory."

YOUR KNOWLEDGE BASE

Think of the information that is stored in your brain as your knowledge base. Your brain stores all kinds of information, including:

- ❑ Dates
- ❑ Statistics
- ❑ Pictures (of people's faces, of places, and of things, for example)
- ❑ Names
- ❑ Addresses
- ❑ Phone numbers

In other words, the total inventory in your brain, your total memory and knowledge, make up your knowledge base.

SELF-CHECK 1-5

Directions: *Write one example for each of the following items:*

1. *Dates* Write your date of birth. ______________________________
2. *Statistics* Write your height and weight.________________________
3. *Pictures* Draw the flag of any country, and write the country's name to the right of it.

   ____

4. *Names* Write the names of two of your friends. ________________

   ____

5. *Addresses* Write your address. ______________________________

What, then, is "learning"? Learning is adding new information to your knowledge base—that is, adding to what you already know.

Stop for a moment to consider what is in *your* knowledge base. Think about one topic—bicycles. Take about five minutes to try to jot down everything you know about bicycles. Use the list that follows as a starting point. This is not a test—just a way of helping you to see how much information your brain has stored away about bicycles.

- ❑ All the experiences you have had riding, borrowing, racing, or repairing bicycles.
- ❑ All you know about bicycle gears—how they work, the names of the various types of bicycles, and what kind of terrain they cover.
- ❑ The names of different bicycle manufacturers, the popular types of bicycle clothing, and so on.
- ❑ Statistics about bicycles—safety, costs, exercise value, use of helmets, and so on.
- ❑ The kinds of bicycles your family, friends, and coworkers own and ride.

SELF-CHECK 1-6

Directions: *Jot down on a separate sheet of paper what you know about three of the following everyday topics. Spend about two minutes on each topic.*

1. Movies
2. Business
3. Finance
4. Sports
5. School
6. Advertising

Now choose *one* of the topics, and write what you know about it. Write three or four sentences in the space provided.

____

____

____

____

Don't be concerned about how much information you have stored away on these topics. Instead, think about how this information entered your brain and became part of your knowledge base. We are concerned with the *process* of learning.

HOW YOUR BRAIN PROCESSES INFORMATION

Learning means adding new information to your knowledge base, adding to what you already know. The brain adds new information to your knowledge base in two different ways: by experience and by practice. Let's take a look at each.

EXPERIENCE. Whenever you see, taste, touch, hear, and talk, you remember some or all of that experience and store it in your knowledge base. For example, when you thought of bicycles earlier, you might have remembered:

1. Seeing a very fancy or unusual bicycle—or a picture of one.
2. Racing on your bicycle against your friends.
3. The feel of the wind on your face when riding your bicycle on a cold day.
4. Hearing a new horn on your brand-new mountain bike.
5. Talking with a friend about a biking trip or your bicycle.

SELF-CHECK 1-7

Directions: *Write a specific example, stored in your knowledge base, for each of the five experiences described above.*

1. ____

2. ____

3. ____

4. ____

5. ____

Each of these experiences is stored in your knowledge base. Your brain will also store other experiences, such as:

1. Seeing the exact words on a page.
2. Tasting a certain food.
3. Touching a soft fabric.
4. Hearing a favorite song or hearing a joke.
5. Talking to a friend or coworker about a certain topic.

No matter how the information gets there, your brain, or knowledge base, stores some information for a short time and some information for a long time. And once it is in your memory, you can use the information to help you learn more. You add to what is already stored in your brain.

Think of what might happen if you tried to drive a motorcycle for the first time. Would your earlier experience on a bicycle be helpful? Most likely it would be. If you were about to use a word processor for the first time, would your past experience on a typewriter be helpful? Sure it would be. In both instances, the "old" experience stored in your knowledge base would help you learn something new.

When you read, you can also use past experiences to help you understand what you are reading. If you are reading about a computer keyboard, you can use your experience with a typewriter keyboard to help you understand basic ideas.

PRACTICE. The more you use a piece of equipment that requires skill, the more you improve your skill. In other words, you learn through practice. For example, hammering nails becomes easier when you've repeated the process a few times. When you travel the same route day in, day out, for several months, the repetition helps you to learn that route.

As you will see, practice also helps you improve your reading skill. New, unknown words become familiar through practice, just as that travel route became familiar through practice.

SELF-CHECK 1-8

Directions: *Name five activities in which you improved your skill through practice.*

1. Activity:______________________________

2. Activity:______________________________

3. Activity:______________________________

4. Activity:______________________________

5. Activity:______________________________

PART 2

Reading to Do

LOCATING INFORMATION IN BOOKS

Books offer tools that help readers find information. The *Construction Trades Knowledge Base* has many of these tools. Working with this book and the *Knowledge Base*, you will learn about the following "finding tools":

- Table of Contents
- List of Figures
- Glossary
- Index

Each of these tools is described below.

THE TABLE OF CONTENTS

The table of contents lists the major topics covered in a book. A portion of the contents from your *Construction Trades Knowledge Base* is shown in Figure 2-1.

Figure 2-1

The table of contents is always in the front of a book. It is usually in the first few pages. Find the *Construction Trades Knowledge Base* Table of Contents. What kind of information does it provide?

As you can see, it offers very useful information. The first part tells you where in the *Construction Trades Knowledge Base* you can find the Preface and the List of Figures.

Notice that the page numbers to the right of *Preface* and *List of Figures* are lowercase roman numerals. That is because they are in a part of the book called the *front matter*. The front matter is numbered separately from the main part of the book. The *Knowledge Base* Table of Contents is also part of the front matter. It is found on pages iii to vi.

On the left side, the contents page shows both the title and the number of each chapter in the *Knowledge Base*. For example:

Chapter 1 Working in the Construction Trades

Below each chapter title, you'll see a list of the subjects discussed in that chapter. For example:

Chapter 3 Designing the Project

(Topic 1) The Architect's Role

(Topic 2) Reading Plans

Just as the chapter titles tell you what's in the book, these subject listings tell you what's in each chapter.

Some subjects are divided into subtopics. Subtopics can also have their own subtopics. For example, look up Chapter 5 in the Table of Contents for the *Knowledge Base*. Chapter 5 is divided into three main topics:

(Topic 1) Finding the Best Location

(Topic 2) Preparing the Site

(Topic 3) Putting in the Foundation

One of these main topics has subtopics. The section "Putting in the Foundation" has three subsections:

(Topic 3) Putting in the Foundation

(Subtopic 1) Footings

(Subtopic 2) Slabs

(Subtopic 3) Foundation Walls

Look at pages 55 and 56 of the *Knowledge Base*. Notice that the main topic, "Putting in the Foundations" appears larger than the subtopics, "Footings" and "Slabs."

Look in the Table of Contents at the section in Chapter 1 called "Jobs in Construction." It has three subtopics:

(Subtopic 1) The Building Trades

(Subtopic 2) Management

(Subtopic 3) Professionals

Each of the three subtopics has its own subtopics. For example, "Professionals" has three subtopics.

(Subtopic 1) Architects

(Subtopic 2) Draftspeople, or Drafters

(Subtopic 3) Engineers

Look at page 13 of the *Knowledge Base*. "Professionals" is underlined. The subtopics to "Professionals," on page 14, which can be referred to as "sub-subtopics," are not underlined. They also appear in smaller type than "Professionals."

The page numbers in the Table of Contents (shown to the right of each chapter title and each section) tell you the first page of that chapter or section so that you can find it quickly.

Chapter 3 Designing the Project *25*

Now spend a few minutes looking over the Table of Contents in your *Construction Trades Knowledge Base*. Then do the following Self-Check.

SELF-CHECK 2-1

Directions: *Use the Table of Contents in the* Knowledge Base *to answer the following questions.*

1. How many chapters are there in the *Knowledge Base*? ____________

2. What is the title of Chapter 6?

____

3. How many main topics are discussed in Chapter 6? ____________

4. Name *all* the topics and subtopics in Chapter 6.

____

____

____

____

____

____

____

____

____

____

____

____

5. On which page does Chapter 2 begin? ____________

6. In which chapter will you find information on:

Wood Exteriors? ____________

Stairs? ____________

7. In a sentence or two, write about how a table of contents can help you locate information. ______________________

____

____

❏ THE LIST OF FIGURES

If a book has many tables, charts, and drawings, it may include a list of figures (illustrations) in the front. Find the page number for the List of Figures in the Table of Contents of your *Construction Trades Knowledge Base*. Turn to that page. Part of the List of Figures is shown in Figure 2-2 below.

Figure 2-2

The List of Figures gives the titles and page numbers of all the figures in the *Knowledge Base*. Like a chapter, each figure has a number and title. If you wanted to see a picture of a site plan, for example, you could use the List of Figures to find out what page it is on—instead of searching the whole book.

Every figure listed has two numbers that identify it. The first number tells you what chapter the figure is in; the second number tells you its order in the chapter. Each figure also has a title that describes its content. For example, in the list of figures, you find this:

Figure 4-4 *Installing wallboard.*

That means that the picture that shows the installation of wallboard is the fourth illustration in Chapter 4.

SELF-CHECK 2-2

Directions: *Use the List of Figures in the* Knowledge Base *to answer the following questions.*

1. How many figures are there in Chapter 6? ____________

2. Give the number of the figure that shows a scale on a drawing.

____

3. On what page will you find Figure 5-2? ____________

4. Which chapter has the most figures? ____________

5. Which chapter has an illustration of a plumber's bid? ____________

6. On what page is the last figure in the *Knowledge Base*? ____________

7. In a sentence or two, write about how a list of figures can help you save time while you read. ____________

____

____

As you have just learned, the *Knowledge Base* contains many figures. It is often easier to find information in a figure than it is to find information in a paragraph or more of text. Look at Figure 3-3 on page 30 of the *Knowledge Base*.

The title appears beneath the illustration and tells you its subject. This figure is a drawing detailing the second floor of a house. By looking at the figure and its title—also called a caption—you learn several things:

❑ You learn that the second floor will have a balcony area open to the living and dining room area below.

❑ You learn the dimensions of each room.

❑ You learn the layout of the whole second story.

❑ You see where the stairway to the first floor fits in the plan.

SELF-CHECK 2-3

Directions: *Use the figures indicated to answer the following questions.*

1. Figure 2-2 on page 17 of the *Knowledge Base*

   a. What is the marital status for the first two employees listed?

   ____

   b. What are the items that need to be filled in to complete the payroll register? ____________

   ____

2. Figure 11-10 on page 111 of the *Knowledge Base*

   a. What is the width of each bookshelf? ____________

   b. How many shelves are shown in the drawing? ____________

   c. How thick is the wood for the shelves? ____________

❑ THE GLOSSARY

A glossary is an alphabetized list of important terms in a book, accompanied by their definitions. Turn to the Table of Contents in the *Knowledge Base,* and find the page on which the Glossary begins. Then skim a few pages of the Glossary to get an idea of what it contains.

You use a glossary the same way you do a dictionary. If you don't understand a term, you can look it up. Remember, the Glossary contains only words or terms related to the *Construction Trades Knowledge Base*. Any words that appear in boldface, or heavy type, in the *Knowledge Base* are defined in the Glossary.

SELF-CHECK 2-4

Directions: *Read the following passage. Use the Glossary in the* Knowledge Base *to answer the questions that follow. (Circle the letter of each right answer.)*

On the second floor of the house, two carpenters are framing the interior walls. They use 2 x 4's as studs. The beams are larger. On the plans, they see four window cutouts. In order to measure exactly, they need to look at the scale of the drawing and find the exact locations of the windows.

As soon as they finish, the electrician and plumber will rough in the spaces needed for the mechanical devices. After inspection, the plans call for blanket insulation covered by a vapor barrier.

1. The term **framing** means:

a. a group of devices that supports a frame

b. all interior parts suspended from the frame

c. the underlying structure of a building without its walls, floors, roof, or ceiling.

2. The term **2 x 4** means:

a. two 4-foot beams

b. a length of pipe

c. a length of lumber approximately 2 inches thick and 4 inches wide.

3. The term **scale** means:

a. a device for removing excess paint

b. a part of the framing system

c. a proportion put on a drawing

❑ THE INDEX

An index is an alphabetical list of topics and key terms discussed in a book, along with the page numbers where they appear. Turn to the Table of Contents in your *Knowledge Base,* and look up the page number of the Index. You will see that it's at the end of the book.

Look at the first page of the Index. As you can see, it has more detailed listings than the Table of Contents. Figure 2-3 shows a portion of the Index.

Figure 2-3

"Construction" is the main entry, or main listing. Subentries are words or terms that relate to the main entry. In this case, the subentries are the items listed under "Construction", such as "types of." Subentries are listed alphabetically below the main entry, with the page numbers on which they can be found. In Figure 2-3, the entry "Cement" is followed by:

See also Concrete

This is a *cross-reference*; it tells you where related information can be found. Below is a different type of cross-reference, called a "see reference":

Homes. *See* Residential construction

It directs you from an entry that is *not used* to one that *is used.* In this example, you are told that any information about homes will be listed under the main entry *Residential construction*.

SELF-CHECK 2-5

Directions: *Use the Index in the* Knowledge Base *to answer the following questions.*

1. What page(s) contain information about framing? ____________

2. How many subentries are there under "Construction"? __________

3. What is the main entry directly before "Wallboard"? __________

4. List the subentries under "Architects."

____

____

____

____

____

5. What pages discuss the topic "Foundation"? ______________
6. In how many places in the *Knowledge Base* is the term *auger* discussed? ______________

PART 2 *Exercises*

THE TABLE OF CONTENTS

1. On what page does Chapter 8 start? ______________
2. In what chapter will you find the topic "Masonry Exteriors"? ______________
3. How many topics, subtopics, and sub-subtopics are there in Chapter 4? ______________
4. What section in Chapter 4 contains information about materials used in construction? ______________
5. On what page does the subtopic "Fasteners and Adhesives" begin? ______________

THE LIST OF FIGURES

1. How many figures are there in Chapter 3? ______________
2. How many figures are there in Chapter 9? ______________
3. On what page is Figure 3-3 located? ______________
4. What is the title of Figure 3-3? ______________
5. On what page is Figure 8-2 located? ______________
6. What is the title of Figure 8-2? ______________
7. In what chapter do you find a figure that shows a multi-use project? ______________

8. What is the number of the figure showing the forms for holding concrete? ______

9. Turn to page 66 and find Figure 6-6.

   a. What is the caption for the figure? ______

   ______

   b. What is in the figure? ______

   ______

   ______

   c. Draw two of the symbols in the figure and tell what they mean?

10. Turn to page 39 and find Figure 4-2.

   a. What is the caption for the figure? ______

   ______

   b. What is in the figure? ______

   ______

THE GLOSSARY

1. Use the Glossary in the *Knowledge Base* to write the definitions for the following terms:

   a. **form** ______

   ______

   b. **lathe** ______

   ______

   ______

   c. **scale** ______

   ______

2. Use the Glossary to answer the following true or false questions.

   Circle *T* for true or *F* for false.

   a. T F An apprentice teaches a trade to beginners.

   b. T F A partition is part of the outside framing.

c. T F The plot plan shows the site for a building.

d. T F A riser is the same as a stringer.

THE INDEX

1. On what page or pages will you find information about each of the following topics:

   a. Stairs ____________________

   b. Foundations ____________________

   c. Truss systems ____________________

   d. Masonry ____________________

   e. Septic systems ____________________

2. For each of the main entries listed, write the subentries:

   a. Asphalt

   ____

   ____

   b. Framing

   ____

   ____

   ____

   ____

   ____

Reading to Learn

USING PQ3R

In Part 1, you learned the difference between *reading to do* and *reading to learn*. Then, in Part 2, you learned and practiced some reading-to-do tasks. Now, in Part 3, you will practice reading to learn.

When your goal is to master information, you are reading to learn. When you are reading to learn, you are an *active* reader. An active reader thinks *before* reading, *during* reading, and *after* reading. To learn to be an active reader, you will use a proven five-step process called "PQ3R":

To be an active reader, do:

P = Preview Before You Read
Q = Question
R = Read During Reading
R = Recode
R = Review After You Read

❑ *PQ3R*: PREVIEW

The first step in reading to learn is to preview the chapter or section you are about to read. When you preview, you skim the material quickly, reading the topic headings and looking at any illustrations, charts, or tables.

Ceilings also come in a variety of finishes. Ceiling height affects heat circulation. It also gives a larger or smaller appearance to a space. The ceiling also affects how sound travels in the room. Conference rooms often have low ceilings made of **acoustical tile** to muffle sounds. A residence may have **cathedral ceilings** to give a spacious feeling to a living room.

Fasteners and Adhesives

Various fasteners hold boards together. They also hold wall, ceiling and floor materials in place. Nails, screws, bolts, and staples are the most common fasteners. Different materials require different fasteners. Exterior fasteners must be rust-proof. Interior fasteners should not show on the outside of the wall. Some tools require the use of certain fasteners to make a job easier. In general, hand tools can secure almost any type of fastener.

In certain situations, adhesives bond two units together. This is particularly true of wood that is used in cabinetry. With adhesives, materials such as wallpaper can be attached without fasteners showing. Also adhesive gives strength to a bond. It is often used along with screws or nails.

Interior Furnishings

Most architects do not make plans that include furniture. They do, however, include all built-ins and appliances in the plans. Built-ins include cabinets, closets, window seats, and shelving. Appliances are usually found solely in the kitchen area. Plumbing and lighting fixtures are usually specified on the plans.

❑ CONSTRUCTION TOOLS

Construction tools are either hand tools or power tools. Hand tools are moved, struck, or turned by hand. Power tools use either electricity or gas for power.

Figure 3-1

For example, as you skim the pages of Chapter 1 of your *Construction Trades Knowledge Base,* you might see the material shown in Figure 3-1. Note that the topic headings are set off from the text. Topic headings are like road signs; they tell you what's coming up. There are three kinds of headings in the *Knowledge Base—main headings, subheadings,* and *sub-subheadings.* These guide you from topic to topic in the material that you are reading. Notice that the main headings are in capital letters and are larger than the subheadings. The subheadings are underlined. Later, you will also see sub-subheadings are not underlined and are slightly smaller than the subheadings.

SELF-CHECK 3-1

Directions: *Turn to Chapter 1 of the* Knowledge Base, *"Working in the Construction Trades," and complete the following exercise.*

1. Write all the topic headings in Chapter 1. Write the main headings at the left margin. Indent the subheadings and the sub-subheadings a few spaces.

2. How many main topics are there in Chapter 1? ________________

3. How many subtopics are there (include *all* subheadings)? ________

4. How many figures are there in Chapter 1? ____________________

❑ PQ3R: QUESTION

After you preview Chapter 1, "Working in the Construction Trades," ask yourself, "What do I already know about this topic?" This question will call to mind the information you already have in your brain, in your knowledge base. Do the same with each main heading and subheading in the chapter or section—ask yourself what you know about each topic. These questions will help you to focus "prior knowledge" on what you are about to read.

For example, look at page 5 in the *Knowledge Base.* When you see the subheading "Commercial Construction," ask yourself, "What do I know about "Commercial Construction?" "What kind of commercial projects have I seen being built in my area?" As you ask yourself such questions, you prepare to learn more about each topic by recalling your prior knowledge.

SELF-CHECK 3-2

Directions: *Listed below are three subheadings from Chapter 1 of the* Knowledge Base. *For each subheading, ask yourself what you know about that topic. Then write the information in the space provided.*

1. Residential Construction

2. Roofers

3. Architects

❑ PQ3*R*: READ

You have previewed Chapter 1 so that you know which topics it covers. You have asked yourself questions so that you can relate the topics to your prior knowledge, or what you already know about each topic. Now you can actively begin to read.

Note the word *actively.* To make sure that you are reading actively, you must:

1. Underline key words as you read. In many books, some of the key words are already emphasized for you—for example, the words may be in **boldface,** in *italics,* or in CAPITAL letters. (As you previewed the chapter, you probably noticed that the key construction words in the *Knowledge Base* are in boldface. These boldfaced words are defined in the Glossary at the end of the book.
2. Think about the questions you asked yourself in the "Question" step. How do your answers relate to the material you are about to read?

SELF-CHECK 3-3

Directions: *Read the section called "Types of Construction," including all of its subsections, starting on page 4. When you finish, close your book and underline the key words in the following paragraph. As you underline, think about whether the words are familiar—do you have prior knowledge of these terms?*

> The four major types of construction include residential, commercial, heavy, and municipal. All four types of construction can include everything from small repairs to multi-use projects. Whether a project is small or large, the first step is usually to get bids from contractors who can do the work. The lowest bid is not always chosen. Sometimes, a contractor with the best reputation and a higher but reasonable bid will win the contract.

As you completed Self-Check 3-3, you were reading *actively.* The goal of reading to learn is to understand what the author is saying and to remember it later on. The first step toward understanding is to grasp the main idea. This, after all, tells you what the author wants to say.

The *main idea* is the most important point in a paragraph or an article. The sentence that contains the core of the main idea is called the *topic sentence.* It is often the first sentence in the paragraph and is followed by details that support, explain, or describe the point. But the topic sentence is not always the first sentence; it may appear anywhere in the paragraph. So, most paragraphs have two basic parts: the *topic sentence* and *supporting details.*

Look at the following paragraph. The first sentence is the topic sentence here. Do you see how it clearly states the main idea?

> **Residential Construction.** The most common type of construction is residential. It includes all projects in which people will live. That is everything from the smallest cabin to the largest apartment complex.

The topic sentence tells you that the most common type of construction is residential. Then the supporting details follow:

1. It includes all projects in which people will live.
2. That is everything from the smallest cabin to the largest apartment complex.

These details support the main idea.

Now read the following example to see how a topic sentence can be placed at the end of a paragraph:

> A design must include the needs of the client, the climate, the surroundings, and the use of the project. In order to understand all the aspects of a project, the designer must have training. He or she must know the building requirements and codes. Architects are licensed designers of projects.

The topic sentence tells you that architects are licensed designers of projects. The rest of the paragraph tells you what designs should include and that the designer must have training.

SELF-CHECK 3-4

Directions: *Read the paragraphs in the* Knowledge Base *that are indicated below. Then write the topic sentence and one or two supporting details for each paragraph.*

1. **Paragraph 3 under "Jobs in Construction" on page 8.**

Topic Sentence: ______________________________

____

Supporting Details: ______________________________

____

____

____

2. **Paragraph 1 under "The Building Trades" on page 8.**

Topic Sentence: ______________________________

____

Supporting Details: ____________________

____

____

____

❑ PQ3*R*: RECODE

The goal of previewing, questioning, and reading is, of course, to remember and use the information you read. Sometimes, though, you will forget the information you've read—that's natural. To help yourself remember what you read, it is important for you to *recode* information immediately after reading.

To *recode* means to express the information in another way. For example, you can recode material by putting it in your own words. This process is called *paraphrasing*. Another way to recode is to make a chart of the information or even to draw a picture. Each of these recoding activities—paraphrasing, making charts, drawing pictures—is covered in this book. Let's begin with paraphrasing.

PARAPHRASING Suppose you repeat a paragraph exactly as you read it in a book. That doesn't mean you *understand* what you're saying. On the other hand, expressing an idea *in your own words* does show understanding. If you did not understand the idea, you would not be able to paraphrase it.

To paraphrase a paragraph, (1) identify the topic sentence, which will tell you the main idea. Then (2) identify the supporting details. And finally, (3) try expressing the main idea and the supporting details *in your own words,* in words that feel natural to you.

Before you try paraphrasing, compare the following paragraph with the suggested paraphrase below it.

If an accident should happen, knowing how to respond immediately can prevent further injury. Never handle someone who has been injured unless you are certain of what to do. Call immediately for help when necessary.

Possible Paraphrase:

You should not move an injured person unless you know how. It is possible to cause more injury. If you don't know exactly what to do, call for help right away.

Your paraphrase will be different because you will use *your own words.* As you can see, paraphrasing is an excellent technique for recoding. The paraphrasing exercises that follow will help you to use this recoding tool to remember what you read.

SELF-CHECK 3-5

Directions: *Read the paragraphs indicated below. Then write a paraphrase of each paragraph.*

1. **Paragraph 4 on page 9.** ______________________________

____

____

____

2. **Paragraph under "Ironworkers and Steelworkers" on page 9.**

____

____

____

❑ PQ3*R*: REVIEW

The final step is to review the material you have read. *Review* means to take another look at, or go over again. There are several effective ways to review what you have read:

1. *List the key points.* Write a list of all the main ideas in the material you have read. Be sure to look at the topic sentences as you develop your list.
2. *Write a summary.* A summary also includes the major points in the chapter or article, but it is not written in list form. Instead, the summary is in paragraph form. Again, be sure to look at all the topic sentences as you develop your summary.
3. *Make a list of the key terms.* In some cases, making this list may be the most helpful way to review what you have just read. By highlighting key terms, you will be reminded of main points.
4. *Skim the chapter again.* Repeating the preview process is another effective way to remember what you have read.

Which technique you use will depend on the type of material you are reading. In any case, you will remember more of what you read when you take time to review it.

SELF-CHECK 3-6

Directions: *Complete the following exercises.*

1. Write five of the key points in Chapter 1.

____

____

____

____

____

____

____

____

2. Write the key terms used in Chapter 1.

3. Write a brief summary—no more than two paragraphs long—of Chapter 1.

Practicing PQ3R With Chapter 2

Read Chapter 2 in the *Knowledge Base.* Use the PQ3R method in the exercises that follow.

PREVIEW

1. What is the title of the chapter?

2. List the major section headings in Chapter 2.

3. List the figures in Chapter 2 by number, and write a two- or three-word summary of each figure caption.

____

____

____

____

____

____

QUESTION

1. Here is the title of Chapter 2 stated as a question:

   What is it like to work on a construction site? Think about the question. Try to answer the question based on what you already know.

   ____

   ____

   ____

2. Change the four major section headings of Chapter 2 into questions.

   ____

   ____

   ____

   ____

3. Answer the questions you wrote in Exercise 2 based on what you already know.

   ____

   ____

   ____

   ____

   ____

   ____

   ____

   ____

READ

Read the section about the general contractor on pages 17 and 18 of the *Knowledge Base.* Then complete the following exercise.

1. Turn back to pages 17 and 18, and read the third and fourth paragraphs under "The General Contractor" again. Then close the book, and fill in the blanks in the paragraph below.

Jordan also sees to it that all work follows the (a)________ plan. The architect's plans appear in a set of (b)_______. The drawings must have the approval of (c)______ ________. They also have the approval of the (d)______ that will own the house. Therefore, Jordan has to follow the plans exactly. The finished project must meet everybody's expectations as set forth in the (e)_______.

2. Read the following paragraph, and underline the key words.

The general contractor submits the bid for a job. If necessary, the general contractor gets bids from subcontractors, such as plumbers or electricians. The bids are based on architect's plans. The plans are shown in a set of drawings. The owner and the zoning officials have to approve the plans before any project starts.

3. Turn to page 18, and read the first paragraph under "Craft or Trade Workers." Identify the topic sentence and list the supporting details.

   Topic Sentence: ______________________________

   ____

   Supporting Details: ______________________________

   ____

   ____

   ____

   ____

4. Paraphrase the paragraph you just read for Exercise 3.

   ____

   ____

   ____

   ____

   ____

   ____

5. Turn to page 20 in the *Knowledge Base*. Read the first paragraph under the heading "Stages of Construction." For that paragraph, list the topic sentence and supporting details.

   Topic Sentence: ______________________________

   ____

   Supporting Details: ______________________________

   ____

RECODE

1. Reread the first three paragraphs under the heading "The General Contractor" on page 17. Paraphrase the main idea of each paragraph.

   a. **Paraphrase of Paragraph 1:** ______________________________

   b. **Paraphrase of Paragraph 2:** ______________________________

   c. **Paraphrase of Paragraph 3:** ______________________________

2. Turn to pages 18, 19, and 20 of the *Knowledge Base,* and read the third paragraph under the heading "Craft or Trade Workers." Paraphrase the main idea in the paragraph.

   Paraphrase of Paragraph 3: ______________________________

REVIEW

Now that you have read Chapter 2, do the following review activity.

1. List the major section headings in Chapter 2 and the key points covered in each section. Use a separate sheet of paper.

Practicing PQ3R With Chapter 3

Read Chapter 3 in the *Knowledge Base* using the PQ3R method.

PREVIEW

1. What is the title of the chapter?

   ____

2. List the major section headings in Chapter 3.

3. List the subsections under the first main heading.

4. List the figures in Chapter 3 by number, and write a short summary of each figure caption.

QUESTION

1. Change the title of the chapter into a question.

2. Think about the question you have written in Exercise 1. Try to answer the question based on what you already know.

3. Change the first major section heading into a question.

4. Ask yourself what you know about the topic and how you might answer this question. Write your answer in the space provided.

5. Change the second major section heading into a question.

6. Ask yourself what you know about this topic, and write your answer in the space provided.

7. Look at Figure 3-6 on page 34. What is pictured in the figure? How would you use what is shown in the figure?

READ

1. Read the following paragraph, and underline the key words.

> The first drawing in a set of plans is usually an outline drawing of the whole project. The other drawings show the details of each floor and room. The drawings are done by a draftsperson. Usually before the drawings are even started, an architect goes over the project with the owner. The architect must also take building codes into account.

2. Turn to pages 26 and 27 in the *Knowledge Base*, and read the first four paragraphs under the heading "The Architect's Role." For each paragraph in the section, write the topic sentence and list supporting details in the space provided.

Paragraph 1

Topic Sentence: ______________________

Supporting Details: ______________________

Paragraph 2

Topic Sentence: ______________________

Supporting Details: ______________________

Paragraph 3

Topic Sentence: ______________________

Supporting Details: ______________________________

____

Paragraph 4

Topic Sentence: ______________________________

____

Supporting Details: ______________________________

____

____

____

____

____

3. Turn to pages 27, 28, and 29 in the *Knowledge Base.* For the first four paragraphs in the section called "Putting the Design on Paper," write the topic sentence and supporting details.

Paragraph 1

Topic Sentence: ______________________________

____

Supporting Details: ______________________________

____

Paragraph 2

Topic Sentence: ______________________________

____

Supporting Details: ______________________________

____

____

Paragraph 3

Topic Sentence: ______________________________

____

Supporting Details: ______________________________

____

____

____

____

Paragraph 4

Topic Sentence: ______________________________

____

Supporting Details: ______________________________

____

____

4. Turn to page 29 of the *Knowledge Base.* For the three paragraphs in the subsection called "Getting Bids," write the topic sentence and supporting details.

Paragraph 1

Topic Sentence: ______________________________

____

Supporting Details: ______________________________

____

____

____

Paragraph 2

Topic Sentence: ______________________________

____

Supporting Details: ______________________________

____

Paragraph 3

Topic Sentence: ______________________________

____

Supporting Details: ______________________________

____

____

____

5. Turn to pages 29, 30, and 31 in the *Knowledge Base,* and find the subsection called "Getting Permits." For the paragraph in the section, write the topic sentence and supporting details.

Topic Sentence: ______________________________

____

Supporting Details: ______________________________

____

____

____

____

____

RECODE

1. Turn to page 31 in the *Knowledge Base,* and reread the first six paragraphs in the section called "Reading Plans." Write a one-paragraph paraphrase of all the information in these paragraphs.

____

____

____

____

____

____

____

____

____

2. Turn to page 34 in the *Knowledge Base,* and read the three paragraphs on that page. Paraphrase each of the three paragraphs in the space provided.

   a. **Paraphrase of Paragraph 1:**

   ____

   ____

   ____

   b. **Paraphrase of Paragraph 2:**

   ____

   ____

   ____

   c. **Paraphrase of Paragraph 3:**

   ____

   ____

REVIEW

1. Write a five-paragraph summary of Chapter 3 on a separate sheet of paper. Recall that a chapter summary is one or more paragraphs that present the main ideas of a chapter. The first paragraph should cover the information in the section called "The Architect's Role." The second paragraph should summarize the subsection "Putting the Design on Paper." The third paragraph should summarize the information in the "Getting Bids" subsection." The fourth paragraph should discuss "Getting Permits," and the fifth paragraph should summarize "Reading Plans."

2. Create a list of the key boldface terms in Chapter 3 on a separate sheet of paper. Define each one.

Practicing PQ3R With Chapter 4

Read Chapter 4 in the *Knowledge Base*. Use the PQ3R method in the exercises that follow.

PREVIEW

1. What is the title of the chapter?

 ____

2. List the major section headings in Chapter 4.

 ____

 ____

3. List the subtopics for the first major section.

 ____

 ____

 ____

 ____

4. List the figures in Chapter 4 by number, and write a two- or three-word summary of each figure caption.

 ____

 ____

 ____

 ____

 ____

 ____

 ____

 ____

 ____

QUESTION

1. Change the title of the chapter into a question.

 ____

2. Think about the question you have written in Exercise 1. Try to answer the question based on what you already know.

3. Change the first major section heading into a question.

4. Ask yourself what you know about the topic and how you might answer this question. Write your answer in the space provided.

5. Change the second major section heading into a question.

6. Ask yourself what you know about this topic, and write your answer below.

7. Look at Figure 4-8. What is pictured in the figure? How does a carpenter use this tool?

READ

1. Read the following paragraph, and underline the key words.

After the project design is approved, the construction work begins. The first step is to bring all tools and materials to the site. There are many construction materials. Concrete and steel are commonly used for foundations and footings. Many homes have wood framing. Large buildings usually have steel-reinforced concrete slabs on every floor. Exterior finishes often depend on the climate.

2. Turn to pages 38, 39, and 40, and find the subsection "Exterior Materials." For the first two paragraphs in the subsection, list the topic sentence and supporting details in the space provided.

Paragraph 1

Topic Sentence: ______________________________

____

Supporting Details: ______________________________

____

____

Paragraph 2

Topic Sentence: ______________________________

____

Supporting Details: ______________________________

____

____

3. Turn to pages 40, 41, and 42, and find the subsection "Interior Materials." For the first four paragraphs in the subsection, list the topic sentence and supporting details in the space provided.

Paragraph 1

Topic Sentence: ______________________________

____

Supporting Details: ______________________________

____

____

____

____

Paragraph 2

Topic Sentence: ______________________________

____

Supporting Details: ______________________________

____

____

____

Paragraph 3

Topic Sentence: ______________________________

____

Supporting Details: ______________________________

____

____

Paragraph 4

Topic Sentence: ______________________________

____

Supporting Details: ______________________________

____

RECODE

1. Reread the paragraphs in the subsection "Fasteners and Adhesives" on page 42. Using the topic sentence and supporting details, paraphrase each paragraph.

   a. **Paraphrase of Paragraph 1:**

   ____

   ____

   ____

   b. **Paraphrase of Paragraph 2:**

   ____

   ____

2. Turn to page 42 in the *Knowledge Base,* and find the subsection called "Interior Furnishings." Paraphrase the information in the paragraph.

   Paraphrase:

   ____

   ____

   ____

3. Turn to page 43 of the *Knowledge Base,* and find the subsection "Hand Tools." Paraphrase the first paragraph in this section.

   Paraphrase:

   ____

   ____

4. Turn to page 46 of the *Knowledge Base,* and find the subsection called "Power Tools." Paraphrase the first two paragraphs in this subsection.

   Paraphrase of Paragraph 1:

   ____

Paraphrase of Paragraph 2:

REVIEW

1. Write a two-paragraph summary of Chapter 4 on a separate sheet of paper. Recall that a chapter summary is one or more paragraphs that present the main ideas of a chapter. The first paragraph should summarize construction materials. The second paragraph should summarize construction tools and safety.

Practicing PQ3R With Chapter 5

Read Chapter 5 in the *Knowledge Base* using the PQ3R method.

PREVIEW

1. What is the title of the chapter:

2. List the major section headings in Chapter 5.

3. List the subsections in the last section.

4. List the figures and figure captions in Chapter 5.

QUESTION

1. Change the title of Chapter 5 into a question.

2. Think about the question you have written in Exercise 1. Try to answer the question based on what you already know.

3. Change the first major section heading into a question.

4. Ask yourself what you know about the topic and how you might answer this question. Write your answer in the space provided.

5. Change the second major section heading into a question.

6. Ask yourself what you know about this topic, and write your answer below.

7. Change the third major section heading into a question.

8. Ask yourself what you know about this topic, and write your answer below.

9. Look at Figure 5-1. What is pictured in the figure? If you have seen such work being done, describe what it was like.

READ

1. Read the following paragraph, and underline the key words.

> The site of a project is its location. Site preparation includes examining the ground for any moisture or other potential problems. It also means leveling the site as called for on the plans. Heavy equipment is often needed to level the site and dig the foundation.

2. Turn to the section "Finding the Best Location" on pages 52 and 53. List the topic sentences and supporting details for the fourth, fifth, and sixth paragraphs in that section.

   Paragraph 4

   Topic Sentence: ______________________

   Supporting Details: ______________________

   Paragraph 5

   Topic Sentence: ______________________

   Supporting Details: ______________________

   Paragraph 6

   Topic Sentence: ______________________

   Supporting Details: ______________________

3. Turn to page 53 in the *Knowledge Base*. For the first three paragraphs in the section called "Preparing the Site," write the topic sentence and supporting details.

   Paragraph 1

   Topic Sentence: ______________________

Supporting Details: ______________________________

____

Paragraph 2

Topic Sentence: ______________________________

____

Supporting Details: ______________________________

____

____

Paragraph 3

Topic Sentence: ______________________________

____

Supporting Details: ______________________________

____

____

____

4. Turn to page 55 in the *Knowledge Base,* and find the subsection called "Footings." For the paragraph in the subsection, write the topic sentence and supporting details.

   Topic Sentence: ______________________________

   ____

   Supporting Details: ______________________________

   ____

   ____

   ____

RECODE

1. Read the four paragraphs in the subsection on pages 56 and 57 called "Slabs." Write a paraphrase of each paragraph.

   a. **Paraphrase of Paragraph 1:**

   ____

   ____

   ____

   b. **Paraphrase of Paragraph 2:**

   ____

   ____

c. **Paraphrase of Paragraph 3:**

____

____

____

d. **Paraphrase of Paragraph 4:**

____

____

2. Turn to page 57 in the *Knowledge Base,* and find the subsection called "Foundation Walls." Paraphrase the first two paragraphs in this section.

   a. **Paraphrase of Paragraph 1:**

   ____

   ____

   ____

   b. **Paraphrase of Paragraph 2:**

   ____

   ____

REVIEW

1. Write a three-paragraph summary of Chapter 5 on a separate sheet of paper. Recall that a chapter summary is one or more paragraphs that present the main ideas of a chapter. The first paragraph should review finding the best location. The second paragraph should summarize preparing the site. The third paragraph should discuss putting in the foundation.
2. Create a list of the key terms in Chapter 5 on a separate sheet of paper. Define each one.

Review Your Knowledge

The following questions can be answered using the material you read in Chapters 1, 2, 3, 4, and 5 of the *Knowledge Base.* Without looking at the *Knowledge Base,* try to answer each of the questions.

CHAPTER 1

1. What is residential construction? _______________

   ____

2. What is an apprentice? _______________

   ____

3. What work do most construction workers do? ________________

4. Name two types of workers who must follow special safety codes and building standards? ____

____

Circle the letter of the correct answer.

5. Your major job is to design projects. You consult with owners. You must know building codes. Also, your design takes the climate, environment, and building use into account. You are:

   a. a builder

   b. an architect

   c. a draftsperson

   d. an engineer

6. Mary is working as a carpenter. She just graduated from high school and this is her first job. She is:

   a. a skilled carpenter

   b. an apprentice

   c. a draftsperson

   d. an office manager

Circle *T* for true or *F* for false.

7. T F All carpenters are skilled workers.

8. T F An owner always takes the lowest contract bid.

CHAPTER 2

1. What is a general contractor? ______________________________

____

2. What is a structural worker? ______________________________

____

3. What is a mechanical worker? ______________________________

____

4. What is the first stage of a construction project? ______________

____

5. What is framing? ______________________________________

____

Circle the letter of the correct answer.

6. A set of plans is:

   a. designed by an architect

   b. drawn by a zoning official

c. only used by the owner

d. all of the above

7. Building inspectors:

a. check all stages of a project

b. are employed by the local government

c. are responsible for approving safe construction methods

d. all of the above

Circle *T* for true or *F* for false.

8. T F Contractors are required to do the work for the amount of an accepted bid.

9. T F Most construction projects require financing.

CHAPTER 3

1. What is an architect?______________________________

2. What is a draftsperson? ______________________________

3. What is a building permit? ______________________________

4. What is a section view?______________________________

5. What is a scale on a drawing?______________________________

Circle the letter of the correct answer.

6. All the sets of plans for a project must be exactly alike because:

a. all bids are based on the plans

b. approval is given to a set of plans

c. all workers use the plans to construct the project

d. all of the above

7. The scale on a drawing is 1/4 of an inch equals 1 foot. A 6-inch sidewall on the drawing will actually be:

a. 24 feet high

b. 6 feet high

c. 12 feet high

d. 36 feet high

Circle *T* for true or *F* for false.

8. T F Electrical outlets do not need to be shown on the plans.
9. T F The architect takes energy conservation into account in each design.

CHAPTER 4

1. What is the most common material used in residential framing?

____

2. What are the two most common materials used to frame larger buildings? ______________________________________

____

3. What are two common exterior sidings? ______________________

____

4. What is wallboard? ______________________________________

____

5. Name four common types of fasteners. ______________________

____

Circle the letter of the correct answer.

6. A hand tool is:
   a. a tool that does not need electricity
   b. a hammer
   c. a screwdriver
   d. all of the above
7. Power tools:
   a. use electricity, gas, or some other form of power
   b. can be stationary or portable
   c. can cut, shape, drill, bore, or fasten
   d. all of the above

Circle *T* for true or *F* for false.

8. T F The most common stationary saw is a portable circular saw.
9. T F Power and hand tools both require knowledge of their use to ensure safety.

CHAPTER 5

1. What is the site? ______________________________________

____

2. What does the structure of a building sit on? ______________________

____

3. What does a surveyor do? ______________________________

____

4. What is a form for a footing? ______________________________

____

Circle the letter of the correct answer.

5. The most common foundation material is:

   a. steel

   b. concrete

   c. wood

   d. brick

6. Before a foundation is poured:

   a. steel beams are put in place

   b. concrete is poured

   c. forms are built

   d. none of the above

Circle *T* for true and *F* for false.

7. T F Wet soil is good for residential construction.

8. T F Usually, the ground must be made level before a foundation is poured.

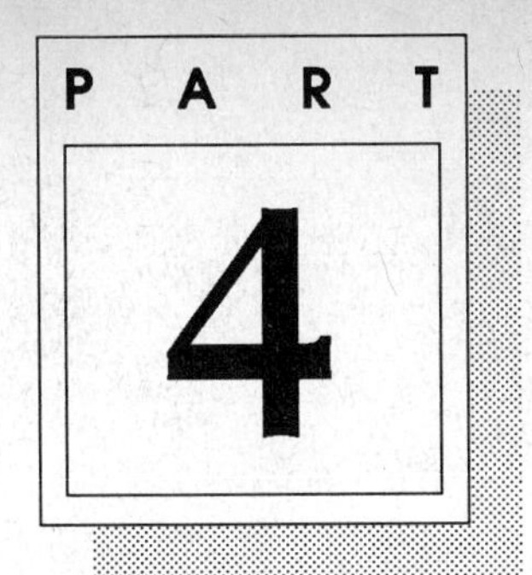

Reading to Learn

MAKING MIND MAPS AND OUTLINES

In Part 3 you learned that one way to remember information that you have just read is to recode it. And one way to recode material is to paraphrase the main ideas—that is, take what you have read and put it into your own words.

❑ Mind Maps

Another way to recode information is to make a mind map. In some ways, a mind map is like a map of a place. Think about what you would do if you were asked where Texas is. You would probably picture a map of the United States in your mind. Then you could look at your mind map to see how and where all the states fit together.

When you make a mind map, you create a picture in your mind of the main idea and supporting details. This picture helps you see, and recall later, how ideas fit together. For example, suppose you have to buy groceries but you forgot to make a list of what you need. You might picture in your mind your local supermarket or grocery store and what products are in each aisle. Then you could go from aisle to aisle in your head, and write items down under categories, such as "Dairy" (milk, eggs, juice); "Produce" (lettuce, apples); "Baking" (flour, sugar); and so on. A mind map of this information would look like Figure 4-1.

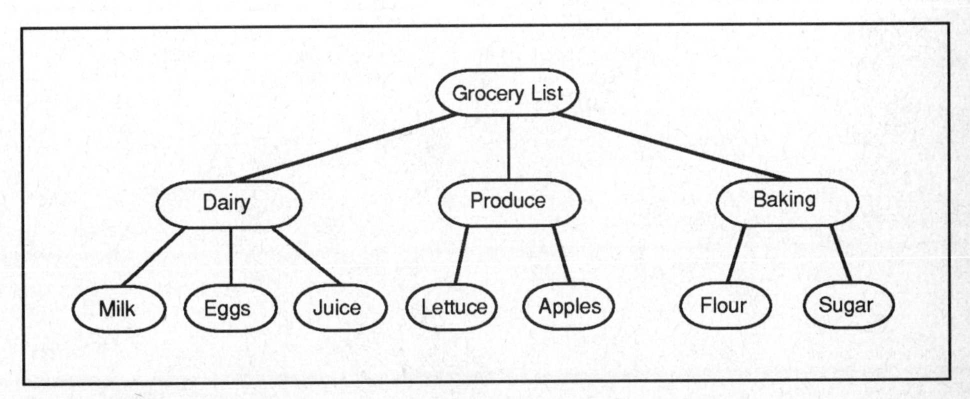

Figure 4-1

You can do the same thing with information that you read and want to remember. You learned in Chapter 2, for example, how work is supervised at a construction site. You learned about the job of general contractor and whom he or she supervises. A mind map of this information would look like Figure 4-2.

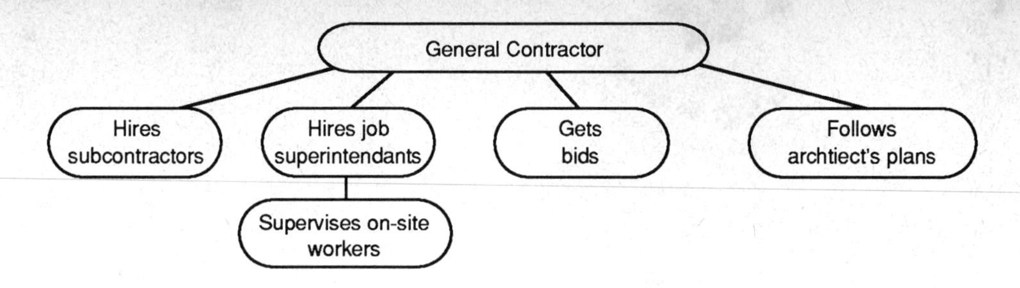

Figure 4-2

SELF-CHECK 4-1

Directions: *Turn to page 51 in the* Knowledge Base, *and read the first paragraph in the subsection entitled "Hand Tools." A mind map for this paragraph has been started for you in Figure 4-3. To finish it, you need to do the following:*

1. Find the four categories of hand tools, and write them in the blank circles beneath the top circle.
2. Draw lines connecting the top circle to those immediately underneath.
3. Write details that support the categories in the circles beneath them.
4. Draw lines from the category circles to the supporting detail circles.
5. Continue adding levels of circles, if necessary, until all of the key information is shown in your mind map.

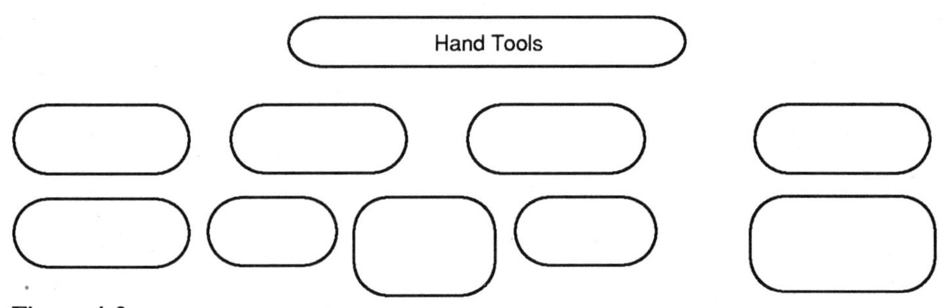

Figure 4-3

Generally, the longer the selection you are mapping, the more levels of circles your mind map will have. A mind map for a few paragraphs is much simpler than one for a whole chapter.

SELF-CHECK 4-2

Directions: *Turn to pages 20 and 21 in the* Knowledge Base, *and read the first two paragraphs in the section called "Stages of Construction." Make a mind map by putting the right information in the circles below.*

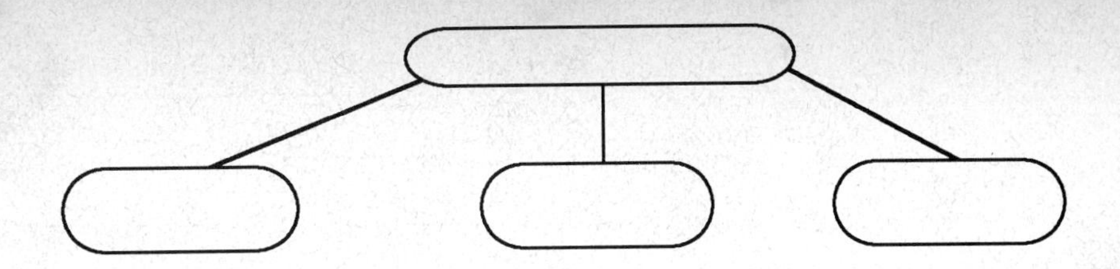

SELF-CHECK 4-3

Directions: *Turn to pages 46, 47, and 48 in the* Knowledge Base, *and read the subsection called "Power Tools." Create a mind map that shows what types of power tools there are and what tasks they perform.*

❑ OUTLINES

Like a mind map, an outline is a way to recode information. Instead of circles connected by lines, however, letters and numbers show how main ideas and supporting details relate to each other. In an outline, the following symbols are used:

- ❑ Roman numerals (*I, II, III, IV, V, VI, VII, VIII, IX, X,* and so on) identify the main ideas.
- ❑ Capital letters (*A, B, C, D,* and so on) identify subtopics that are part of the main idea.
- ❑ Arabic numbers (*1, 2, 3, 4,* and so on) identify details that support subtopics.
- ❑ Lowercase letters (*a, b, c, d,* and so on) identify minor details.

All of these symbols are set up so that supporting details are indented under more important points. For example:

I. Main Idea
    A. Subtopic
        1. Supporting detail
        2. Supporting detail
            a. Minor supporting detail
            b. Minor supporting detail

B. Subtopic
   1. Supporting detail
   2. Supporting detail
II. Main Idea
   A. Subtopic
   B. Subtopic
      1. Supporting detail
      2. Supporting detail
   C. Subtopic

The following passage is based on material from Chapter 1 in the *Knowledge Base.* After the passage, you'll find the same information in outline form. Notice that you do not have to use the exact words from a passage when presenting the information in outline form. In other words, when you make an outline, you often paraphrase.

> There are four major types of construction. The most common portion, residential, includes the construction of all properties meant to be lived in. The next largest, commercial, includes all properties meant for business or public use. Heavy construction includes the building and repair of big projects, such as roads, bridges, and railroads. Municipal construction includes all essential public services projects, such as sewers or sidewalks.

I. Types of Construction
   A. Residential
      1. Most common
      2. Homes
   B. Commercial
      1. Second largest part
      2. Business properties
      3. Public buildings
   C. Heavy construction
      1. Biggest projects
      2. Roads, bridges, dams, and so on
   D. Municipal construction
      1. Public service projects
      2. Sewers, sidewalks
II. Next Main Idea

Please note that, for every level of an outline, you should have at least two items. Therefore, if you label an item "A," you must have at least one more item (labeled "B") at the same level.

SELF-CHECK 4-4

Directions: *Turn to pages 8 in the* Knowledge Base, *and read the first three paragraphs in the section called "Jobs in Construction." (Stop reading when you get to the subsection called "The Building Trades.") Next, look at the following outline for the first part of what you just read. It identifies the* *main idea next to roman numeral* I. *It also identifies the first topic of the main idea next to the letter* A *and the two details that support topic* A. *Continue the outline by writing the next main topic beside the letter* B, *then writing supporting details. Do the same with the letter* C.

I. Jobs in Construction
   A. The building trades
      1. Unskilled workers
      2. Craft workers
   B. ______________________________
      1. ______________________________
      2. ______________________________
   C. ______________________________
      1. ______________________________
      2. ______________________________
II. Next Main Idea

SELF-CHECK 4-5

Directions: *Read the material in the* Knowledge Base *beginning with the section called "The Job Superintendent" on page 18. (Stop when you get to the next section head, "Craft or Trade Workers. Based on your reading, create an outline for this material on a separate sheet of paper.*

Practicing PQ3R With Chapter 6

PREVIEW

1. What is the title of Chapter 6? ______________________________

   ____

2. List all of the major topic headings in Chapter 6.

   ____

   ____

   ____

   ____

3. List all the subtopics under the first major topic heading in Chapter 6.

   ____

   ____

4. List the figures and figure captions in Chapter 6.

____

____

____

____

____

____

____

____

QUESTION

1. In the space provided, write what you already know about each of the major topic headings in Chapter 6.

   a. ____

   ____

   ____

   b. ____

   ____

   ____

   c. ____

   ____

   ____

   d. ____

   ____

   ____

READ

1. Turn to page 60, 61, and 62 in the *Knowledge Base.* Read the section called "Framing the Floor." When you have finished, close the book and fill in the blanks in the following paragraph.

In the foundation, workers placed (a)_______ _____. The anchor bolts hold the sill (b) _____. The (c)_______ code requires a metal termite (d)_____.

2. What is the main idea in the preceding paragraph?

____

____

3. Read the subsection called "Subflooring" on page 61 of the *Knowledge Base*. For each paragraph in this section, write the topic sentence in the space below.

   Paragraph 1

   Topic Sentence: ______________________________

   ____

   Paragraph 2

   Topic Sentence: ______________________________

   ____

4. Read the subsection in the *Knowledge Base* called "Finishing the Floor" on page 62. When you are finished, close your book and answer the following true-false questions. **(Circle *T* for true or *F* for false.)**

   a. T F The subfloor is usually made of carpeting.

   b. T F Masons complete all framing work.

   c. T F The subfloor is suitable for holding many types of finished flooring.

5. Turn to pages 62, 63, and 64 in the *Knowledge Base*, and find the section entitled "Framing the Walls." Read the first three paragraphs in that section. In the space below, restate in your own words the main idea for each paragraph.

   Paragraph 1

   Main Idea: ___________________________________

   ____

   Paragraph 2

   Main Idea: ___________________________________

   ____

   Paragraph 3

   Main Idea: ___________________________________

   ____

   ____

6. Read the subsection called "Inside Walls," on page 65 of the *Knowledge Base*. When you have finished, close your book and fill in the blanks in the following paragraph.

The inside walls are called (a)________. The wall framing consists of (b)______ attached to a top and bottom board. The mechanical devices must be (c)_______ __ before the wallboard is placed over the frame. Some walls have (d)________ for doors, built-in shelving, pass-throughs, and closets.

7. Turn to pages 65, 66, and 67 of the *Knowledge Base*, and read the subsection called "Plumbing." In the space below, restate in your own words the main idea for each paragraph in that section.

   Paragraph 1

   Main Idea: ____

   ____

   Paragraph 2

   Main Idea: ____

   ____

   Paragraph 3

   Main Idea: ____

   ____

8. Read the subsection called "Electrical Work" on page 67 of the *Knowledge Base*. When you have finished, close your book and fill in the blanks in the following paragraph.

The electrician (a)______ _______ all lines and outlets after the framing stage. The electrician does not bring (b)___________ to the outlets until the finishing stage. All electrical work is approved by (c)_________ before the walls are closed up.

9. Read the subsection called "Ceilings" on pages 67 and 68 of the *Knowledge Base*. Close your book and answer the following true-false questions. **(Circle *T* for true or *F* for false.)**

   a. T F The ceiling frame is similar to a floor frame.

   b. T F On the attic side of the ceiling, it is important to finish the floor.

10. Turn to pages 68, 69, and 70 of the *Knowledge Base*, and find the subsection called "Roof Framing." Read until the end of the chapter. Close your book, and fill in the blanks in the paragraph below.

Most roofs fall into one of the (a)___ basic categories of roof design. (b)_____ roofs have two sloped sides that meet at the top. A (c)___ roof has four sloped sides that meet at the top. A gable-style roof with each slope broken into two or more slopes is called a (d)_______ roof.

RECODE

1. Turn to pages 62 through 65 in the *Knowledge Base,* and find the main section called "Framing the Walls." Draw a mind map for the information in that section, including all its subsections.

2. Turn to pages 68, 69, and 70 of the *Knowledge Base,* and create a mind map of the information in the subsection called "Roof Framing."

REVIEW

1. Create a list of key words in Chapter 6. Define each one.
   Use a separate sheet of paper.

Practicing PQ3R With Chapter 7

PREVIEW

1. What is the title of the chapter? ______________________

   ____

2. List the major topic headings in Chapter 7.

   ____

   ____

   ____

   ____

3. List the figures and figure captions in Chapter 7.

   ____

   ____

   ____

   ____

   ____

   ____

QUESTION

1. Ask yourself what you know about each of the four major topic headings in Chapter 7, and write your responses in the space provided.

   a. ______________________

   ____

   ____

   ____

   b. ______________________

   ____

   ____

   ____

   c. ______________________

   ____

d. ______

READ

1. Turn to pages 72 and 73 in the *Knowledge Base*. Read the section called "Masonry Exteriors." For each of the first three paragraphs, restate in your own words the main idea.

   Paragraph 1

   Main Idea: ______

   Paragraph 2

   Main Idea: ______

   Paragraph 3

   Main Idea: ______

2. Read the section called "Wood Exteriors" on pages 73, 74, and 75. Identify the topic sentence and supporting details for each of the first two paragraphs.

   Paragraph 1

   Topic Sentence: ______

   Supporting Details: ______

   Paragraph 2

   Topic Sentence: ______

   Supporting Details: ______

3. Read the sections called "Aluminum or Vinyl Siding" and "Cement and Stucco Exteriors" on pages 75 and 76 of the *Knowledge Base*. Then close your book, and fill in the blanks in the following paragraph.

A popular exterior material is (a)____ ______, which is a manufactured product. As (b)___ becomes scarcer, vinyl is becoming more popular. In warm climates, (c)_____ or (d)______ exteriors stand up to intense sun. Stucco is spread over a wire (e)____.

RECODE

1. Turn to pages 72 and 73 in the *Knowledge Base.* Reread the section called "Masonry Exteriors." Write a one-paragraph paraphrase of all the information in that section. ____________________

   ____

   ____

   ____

2. Turn to pages 73, 74, and 75 in the *Knowledge Base.* Restate in your own words the main idea for the third, fourth, and fifth paragraphs under the heading "Wood Exteriors."

   Paragraph 3

   Main Idea: ____________________________________

   ____

   Paragraph 4

   Main Idea: ____________________________________

   ____

   Paragraph 5

   Main Idea: ____________________________________

   ____

3. Make a mind map of the section titled "Wood Exteriors" on pages 73, 74, and 75.

REVIEW

1. Create a list of key terms in Chapter 7, using a separate sheet of paper. Then define each one.

Practicing PQ3R With Chapter 8

PREVIEW

1. List the section headings in Chapter 8.

____

____

____

2. How many figures are there in Chapter 8? ______________
3. List the figures and figure captions in Chapter 8.

____

____

____

____

____

____

____

QUESTION

1. Ask yourself what you know about the three section headings in Chapter 8. Write your response in the space provided.

a. ______________________________

____

____

b. ______________________________

____

____

c. ______________________________

____

____

READ

1. Turn to pages 78 and 79 in the *Knowledge Base.* Read the third and fourth paragraphs under the heading "Insulation." Identify the topic sentence in each paragraph.

   Paragraph 3

   Topic Sentence: ____

   ____

   Paragraph 4

   Topic Sentence: ____

   ____

2. Read the section called "Insulation" on pages 78 and 79 of the *Knowledge Base.* Then close your book and fill in the blanks in the following paragraph.

> Reflective insulation has a (a)____ surface that reflects (b)____. All insulation needs some kind of moisture or (c)_____ ________ to protect it from (d)____________.

3. Turn to page 83 of the *Knowledge Base,* and read the section called "Ceilings." Write the topic sentence and supporting details for each paragraph in this section.

   Paragraph 1

   Topic Sentence: ____

   ____

   Supporting Details: ____

   ____

   Paragraph 2

   Topic Sentence: ____

   ____

   Supporting Details: ____

   ____

   ____

   Paragraph 3

   Topic Sentence: ____

   ____

Supporting Details: ______________________________

____

RECODE

1. Turn to pages 78 and 79 in the *Knowledge Base,* and reread the section called "Insulation." Then create a mind map of the information.

2. Read the section called "Walls" on pages 79 through 83 of the *Knowledge Base.* Write a one-paragraph paraphrase of all the information in this section.

____

____

____

____

____

____

3. Turn to page 83 in the *Knowledge Base.* Write an outline of the information in the section called "Ceilings."

____

____

____

____

____

____

REVIEW

1. On a separate sheet of paper, create a list of the key terms, in Chapter 8. Then define each one.

Practicing PQ3R With Chapter 9

PREVIEW

1. What is the title of Chapter 9? ______________________________

   ____

2. List the two major topic headings in Chapter 9.

   ____

   ____

3. List the figures and figure captions in Chapter 9.

   ____

   ____

   ____

   ____

   ____

   ____

   ____

   ____

QUESTION

1. Ask yourself what you already know about each of the two main section headings in Chapter 9 of the Knowledge Base. Write what you know in the space below.

   a. ____

   ____

   ____

   b. ____

   ____

   ____

READ

1. Turn to pages 86 to 89 of the *Knowledge Base.* Read the section called "Stairs." In the space provided below, write a one-paragraph paraphrase of the main ideas in the first three paragraphs in that section.

____

____

____

____

2. Reread the last two paragraphs in the "Stairs" section on page 89 of the *Knowledge Base.* When you have finished, close your book and fill in the blanks in the following paragraph.

> Outside steps have to be (a)____________. (b)_________ often forms the base to avoid insect damage. Disappearing stairs provide (c)_______ to attic and other storage areas. These stairs (d)____ __ into a covered opening in the ceiling.

3. Read the first paragraph in the section called "Floors" on page 89 of the *Knowledge Base.* In the space below, write the topic sentence and supporting details for that paragraph.

Topic Sentence: ____

____

Supporting Details: ____

____

____

____

____

RECODE

1. Turn to page 92 of the *Knowledge Base.* Read the last paragraph in Chapter 9. Write a paraphrase of that paragraph.

____

____

____

2. Reread pages 86 to 89 of the *Knowledge Base.* Draw a mind map of all the information in that section on the rules of stair design.

3. Make an outline of the information in the section called "Floors," on pages 89 through 93 of the *Knowledge Base.*

REVIEW

1. On a separate sheet of paper, create a list of key terms in Chapter 9. Define each one.
2. Write a summary paragraph for each of the two major sections in the chapter. Use a separate sheet of paper.

Review Your Knowledge

The following questions can be answered using the material you read in Chapters 6, 7, 8, and 9. Without looking at the *Knowledge Base,* try to answer each question.

CHAPTER 6

1. What holds the frame to the foundation? ____________________

____

2. What is a post? ______________________________________

____

3. What is a girder? ____________________________________

____

4. What is the difference between the subfloor and the finished floor?

____

____

Circle the letter of the correct answer.

5. Plan drawings show:
   a. approximate measurements
   b. about where the windows will be placed
   c. the placement of furniture
   d. exact to-scale placement of windows and doors

Circle *T* for true or *F* for false.

6. T F Double studs help make corners stronger.
7. T F The openings for windows are cut out of the finished walls.
8. T F Electricians and plumbers rough in mechanical devices after the framing stage.
9. T F Building inspectors only check the final stages.
10. T F There are six basic roof designs.

CHAPTER 7

1. What is a veneer? ____________________________________

____

2. What are three types of wood siding? ____________________

____

3. Give two reasons why vinyl siding is popular. ______________

____

Circle the letter of the correct answer.

4. Vinyl siding is:

   a. installed with relative ease.

   b. a manufactured product

   c. available in a variety of colors

   d. all of the above

Circle *T* for true or *F* for false.

5. T F Many stone and brick exteriors are really veneers.

6. T F Wood siding is rarely used.

7. T F Weathered wood is painted.

8. T F Stucco and cement are rarely used in warm climates.

CHAPTER 8

1. What is an R-value? ____________________

   ____

2. What is blanket insulation? ____________________

   ____

3. Name three other types of insulation? ____________________

   ____

4. What is drywall? ____________________

   ____

5. Define the following terms:

   a. STC rating: ____________________

      ____

   b. vapor barrier: ____________________

      ____

Circle the letter of the correct answer.

6. Ceramic tile:

   a. is relatively expensive to install

   b. requires a lot of maintenance

   c. is used most often in living rooms

   d. all of the above

Circle *T* for true or *F* for false.

7. T F The drywall covers the framing, insulation, and mechanical devices.

8. T F Ceramic tiles make a good ceiling material.

CHAPTER 9

1. What is a baluster? ______________________________

____

2. What is a riser? ______________________________

____

3. What is a tread? ______________________________

____

4. What is subflooring? ______________________________

____

5. What is tongue-and-groove flooring? ______________________________

____

6. Name two types of subflooring. ______________________________

Circle the letter of the correct answer.

7. When designing stairs:
   a. risers are always larger than treads
   b. risers should never be too high for safety's sake
   c. treads should be twice as wide as risers
   d. there should be no more than 8 steps

Circle *T* for true or *F* for false.

8. T F Outside steps need to be weatherproof.
9. T F The surface of a floor is usually made flat in the finishing stage.

CHANGING TEXT TO PICTURES, TABLES, AND FLOWCHARTS

You have learned to recode information by paraphrasing, and by making mind maps and outlines. Drawing a picture about what you've read is yet another way to recode material. Maybe you've heard the saying, "A picture is worth a thousand words." That saying is based on the belief that ideas are often easier to understand—and remember—when they are presented as a picture instead of in words.

For example, look at the picture of the construction site in the front of the *Knowledge Base.* Because this picture is the first thing you see, it sets the scene for the whole book. When you finish the Knowledge Base and try to recapture the main points, the picture will provide you with a mental image of the work taking place at the construction site.

❑ CHANGING TEXT TO PICTURES

You don't have to be an artist to be able to change text to pictures. To recode information in picture form, you first list the main ideas and supporting details. Then illustrate them in a way that has meaning for you. Later, when you need to recall the information, the picture you drew will help jog your memory about other details.

Turn to pages 4 to 8 of the *Knowledge Base,* and read the section called "Types of Construction." A picture of this information might look something like the one shown in Figure 5-1. Note that the key parts are labeled.

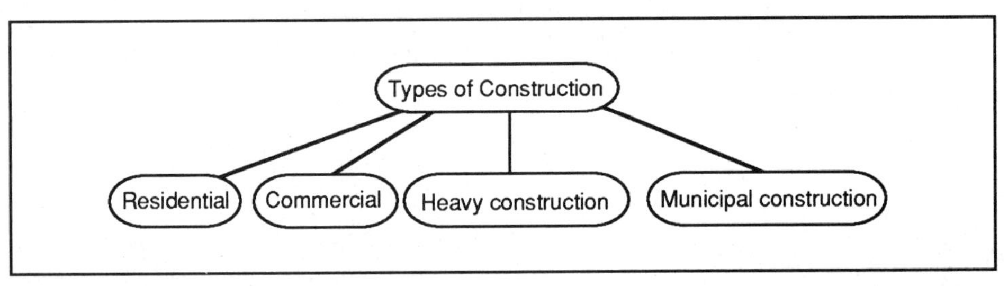

Figure 5-1

SELF-CHECK 5-1

Directions: *Now turn to page 43 in the* Knowledge Base, *and read the paragraph under the subtopic "Hand Tools."*

1. Write the main idea and supporting details for the paragraph.

   Main Idea: ____

   ____

   Supporting Details: ____

   ____

   ____

   ____

   ____

2. In the space provided, recode the information about hand tools as a picture. Remember to label each part of the picture.

❑ CHANGING TEXT TO TABLES

Some types of information lend themselves to recoding as tables. When you have groups of material, a table helps you find specific information quickly. And when you want to recall details, picturing the rows and columns of the table in your mind is a memory aid.

To create a table, you must first put the information in groups or categories. For example, if you were making a table containing information about materials of construction, you might use categories such as "Product," "Use," and "Relative Cost." Then you decide on the best way to present the information. Sometimes, you will use labels only for the *columns* in the table. Sometimes, you will use labels for both *rows* and *columns*.

Turn to pages 46 through 48 of the *Knowledge Base* and find the subsection called "Power Tools." Below is a table that shows a different way to recode some of the information on power tools.

Power Tools

Name of Tool	*Portable*	*Stationary*	*Category of Use*
Circular Saw	x		Cutting
Radial-Arm Saw		x	Cutting
Electric Drill	x		Drilling
Power Plane	x		Cutting
Portable Router	x		Drilling
Portable Sander	x		Cutting
Power Stapler	x		Fastening
Power Nailer		x	Fastening

Each of the places where a row intersects with a column is called a cell. For example, the entry "Fastening" is a cell. It is the cell where the column labeled "Power Stapler" intersects with the row labeled "Category of Use." Now that the material in the sections has been recoded in the form of a table, you can use it to find information quickly.

SELF-CHECK 5-2

Directions: *Turn to pages 68, 69, and 70 in the* Knowledge Base. *Read the subsection called "Roof Framing." Recode the information into the table below.*

Roof Framing

Type of Roof	*Major Characteristic*
Gable	
Hip	
Gambrel	
Flat	
Shed	
Mansard	

SELF-CHECK 5-3

Directions: *Read the paragraph below. After you have finished reading, use the space provided to recode the information in the form of a table.*

There are four major types of construction—residential, commercial, municipal, and heavy. Residential construction makes up the largest segment of all construction projects. It includes the building of all living spaces, such as homes and apartments. Commercial construction is the second most common type of construction. It includes all offices and factories. Next is heavy construction, which includes large public projects, such as bridges and dams. Last is municipal construction, which includes all public works projects, such as sewers and sidewalks.

❑ CHANGING TEXT TO FLOWCHARTS

A flowchart is a diagram that shows each step of a process or system. Recoding information in the form of a flowchart is helpful when you are trying to understand and remember the steps in a process or a series of events that lead to a specific result.

Flowcharts use connecting lines and special symbols, which are described below.

A *circle* indicates the beginning or end of a process or series.

A *box* contains one step in the process.

A *diamond* contains a question; it tells you a decision must be made.

An *arrow* shows the order of steps or events; it connects the other symbols.

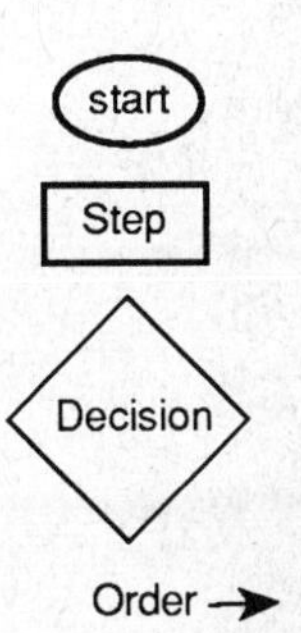

In the two examples that follow, material is given in text form, and then it is recoded on flowcharts.

EXAMPLE 1

To dial a long-distance number direct, you first dial "1." Then you dial the area code followed by the seven-digit local number.

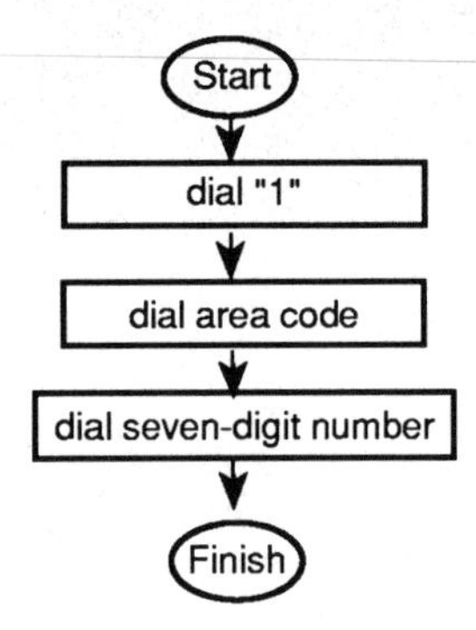

As you can see, the flowchart in Example 1 shows three steps. Below is a different kind of flowchart. It involves a decision.

EXAMPLE 2

When an architect designs a house, the owner takes part in the basic decisions. The owners in this example have stated that either a colonial or a ranch would be fine. The architect will make the decision based on the number of rooms needed. If the family needs more than four bedrooms, the architect will design a colonial.

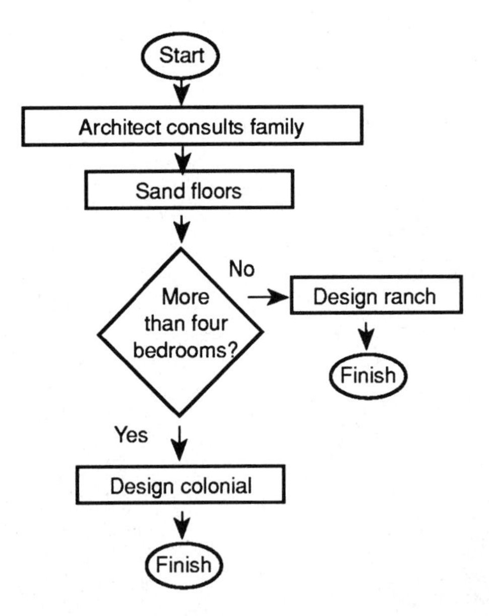

As you can see, when you use a decision box, you create two paths for the flowchart. In Example 2, if the family needs more than four bedrooms, the architect will design in a certain way. If the family needs less than four bedrooms, a ranch house will be designed.

SELF-CHECK 5-4

Directions: *Turn to page 17 and 18 in the* Knowledge Base, *and read the section called "The General Contractor." Recode the information in the flowchart provided here.*

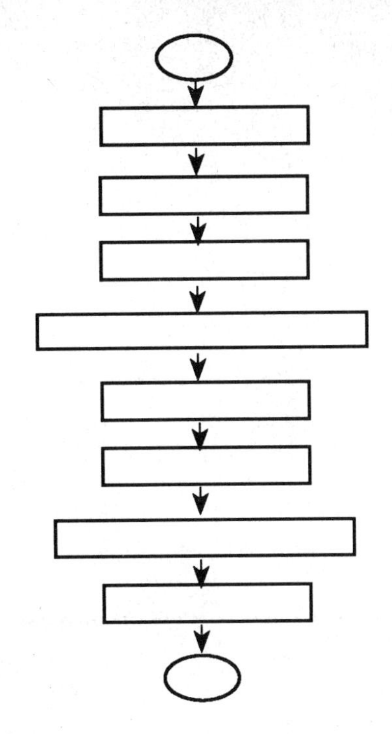

SELF-CHECK 5-5

Directions: *Turn to page 12 in the* Knowledge Base, *and read the first three paragraphs in the subsection called "Management." In the flowchart provided here, recode the information in those paragraphs.*

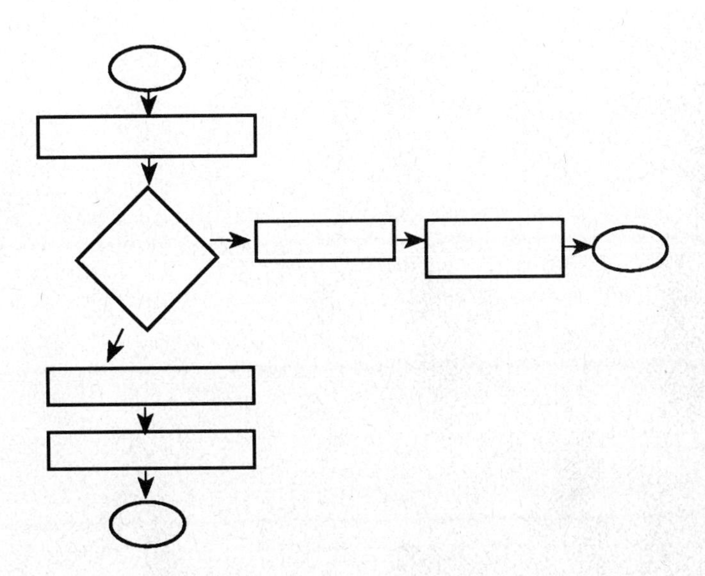

SELF-CHECK 5-6

Directions: *Read the following passage. In the space provided, draw a flowchart that represents the information in the passage.*

The electrician at the Bardel and Baker site has just finished wiring the two-family house. He has checked the building codes, but the special wiring called for is unusual and may require some changes. The building inspector comes. If approved, the carpenters will come in and close up the walls. If not approved, the electrician will make changes and have them inspected. When approved, the carpenters will close up the walls.

Practicing PQ3R With Chapter 10

PREVIEW

1. What is the title of Chapter 10? ______________________________

____

2. List the chapter's section headings.

____

____

____

____

3. List the figures and the figure captions in Chapter 10.

____

____

____

____

____

____

QUESTION

1. For each major section heading in Chapter 10, ask yourself what you already know about the topic. Write your response in the space provided.

   a. ______________________________

   b. ______________________________

   c. ______________________________

   d. ______________________________

READ

1. Turn to page 96 in the *Knowledge Base.* Read the first two paragraphs in the section called "Electricity." Write the topic sentence and supporting details for each of the paragraphs.

   Paragraph 1

   Topic Sentence: ______________________________

   Supporting Details: ______________________________

Paragraph 2

Topic Sentence: ______________________________

____

Supporting Details: ______________________________

____

____

2. Read the first two paragraphs in the section called "Plumbing," on page 98. List the topic sentence for each of the two paragraphs.

Paragraph 1

Topic Sentence: ______________________________

____

Paragraph 2

Topic Sentence: ______________________________

____

3. Turn to pages 111 and read the first three paragraphs on the page. Then close your books, and fill in the blanks in the following paragraph.

Cold-water (a)_____ bring water for toilets, (b)_________, (c)________, sinks, and washing machines to the appropriate points. Hot-water pipes bring hot water from a (d)___-_____ ____ to the appropriate points. Waste pipes remove (e)_________ from the bathrooms and kitchens.

4. Turn to page 101 and 102 in the *Knowledge Base,* and read the section called "Heating." Restate in your own words the main ideas of each of the last two paragraphs in this section.

Paragraph 4

Main Idea: ______________________________

____

Paragraph 5

Main Idea: ______________________________

____

RECODE

1. Turn to pages 99 and 100 in the *Knowledge Base,* and reread the paragraphs on those pages that discuss cold-water, hot-water, and wastewater pipes. Put the information about these pipes in outline form.

2. Read the section called "Air Conditioning" on page 102 of the *Knowledge Base.* Write a three-sentence paraphrase of the information in that section.

3. Reread the paragraphs about pipes on pages 99 and 100. Draw a flowchart of the types of pipes and where they go.

REVIEW

1. Create a list of key terms in Chapter 10 on a separate sheet of paper, and define each one.

Practicing PQ3R With Chapter 11

PREVIEW

1. What is the title of Chapter 11? ______________________________

   ____

2. List the major section headings in the chapter.

   ____

   ____

   ____

   ____

   ____

QUESTION

1. For each major topic heading in Chapter 11, ask yourself what you already know about the subject. Write your response in the space provided.

a. ______________________________

____

____

b. ______________________________

____

____

c. ______________________________

____

____

d. ______________________________

____

____

e. ______________________________

____

____

READ

Turn to pages 104 and 105 in the *Knowledge Base*. Read the section called "Adding Interior Trim."

1. What is the topic sentence in the first paragraph?

   Topic Sentence: ______________________________

   ____

2. Write the topic sentences for the first two paragraphs in the section called "Fixtures and Appliances on pages 105 and 106."

   Paragraph 1

   Topic Sentence: ______________________________

   ____

   Paragraph 2

   Topic Sentence: ______________________________

   ____

3. Close your book, and fill in the blanks in the following paragraph:

> In the framing stage, all pipes were brought to the joining point. Now the plumber is ready to hook up (a)________, such as sinks, toilets, and tubs. For every part of the house that will need water, (b)____-___ valves are installed. In case of (c)_________, the valves can stop the flow of (d)_____ into the area.

Read the section called "Lighting" on page 107.

4. What is the main idea in the first paragraph of this section? The second paragraph? The third paragraph?

Paragraph 1

Main Idea: ______________________________

____

Paragraph 2

Main Idea: ______________________________

____

Paragraph 3

Main Idea: ______________________________

____

5. Turn to page 112 of the *Knowledge Base.* Read the last paragraph on that page, and write the topic sentence and supporting details for that paragraph.

Topic Sentence: ______________________________

____

Supporting Details: ______________________________

____

____

____

RECODE

1. Turn to pages 105 and 106 in the *Knowledge Base,* and reread the section about fixtures and appliances. Draw a picture that represents the information in that section.

2. Turn to page 107 of the *Knowledge Base.* Reread the section called "Lighting." Make on outline that gives the information in that section.

3. Turn to pages 109, and read the first full paragraph on that page. Make a flowchart of the information in that paragraph.

REVIEW

1. List and define all the key terms in Chapter 11 on a separate sheet of paper.
2. Write a short paragraph that summarizes the information in Chapter 11 about the finishing of surfaces. Name the major types of finishes and where they might be used. Use a separate sheet of paper.

Practicing PQ3R With Chapter 12

PREVIEW

1. What is the title of Chapter 12? ____________________

____

2. List the chapter's major section headings.

____

____

____

____

____

3. List the figures in the chapter.

____

____

____

____

____

QUESTION

For each major topic heading in Chapter 12, write what you already know about the subject in the space provided.

1. ____________________

____

____

2. ____________________

____

____

3. ____________________

____

____

4. ____________________

____

____

5. ____

____

____

READ

1. Turn to page 114 in the *Knowledge Base.* Read the section called "Patios and Walkways." Write a one-paragraph summary of the first two paragraphs in that section. ______________________________

____

____

____

____

____

____

____

2. Read the section called "Driveways" on pages 114 and 115 of the *Knowledge Base.* Restate in your own words the main ideas in that section. ______________________________________

____

____

____

3. Read the section entitled"Planting Trees and Shrubs," on pages 115 and 116 of the *Knowledge Base.* Close your book, and fill in the blanks in the following paragraph:

> The landscaper only puts in the main (a)_____ and the (b)____. The home owners intend to add their own flower and vegetable gardens. Bardel and Baker tried to preserve as many (c)_____ as possible. The landscaper takes into account the look of the house, the (d)______, and the (e)_______ of daily sunlight.

4. Read the section called "Lawns," on pages 116 and 117 of the *Knowledge Base.* Write the topic sentence for the last paragraph in that section.

   Topic Sentence: ______________________________

   ____

RECODE

1. Turn to pages 116 and 117. Reread the section called "Lawns." Create a mind map of the information in this section.

2. Turn to page 114 in the *Knowledge Base*, and reread the section called "Patios and Walkways." Create a flowchart showing the steps used in laying down a brick walkway.

REVIEW

1. On a separate sheet of paper, write a brief summary of Chapter 12.

Practicing PQ3R With Chapter 13

PREVIEW

1. What is the title of Chapter 13? ______________________________

   ____

2. List the figures and figure captions in Chapter 13.

   ____

QUESTION

1. List what you already know about the subject of Chapter 13.

   ____

   ____

   ____

   ____

READ

1. Turn to page 120 in the *Knowledge Base.* Read the next to last paragraph. Restate in your own words the main idea of the paragraph.

   ____

   ____

RECODE

1. Turn to pages 119 and 120 of the *Knowledge Base,* and read the entire chapter. Create a flowchart for safety in the construction industry.

2. Outline the entire chapter.

____

____

____

____

____

____

____

____

____

____

____

____

____

____

REVIEW

1. Write a brief summary of Chapter 13 on a separate sheet of paper.

Review Your Knowledge

The following questions can be answered using the information you read in Chapters 10, 11, 12, and 13. Without looking at the *Knowledge Base,* try to answer each question.

CHAPTER 10

1. At what stage are the electrical lines brought to the rooms of the house? ____
2. What is a circuit breaker box? ____

   ____

   ____
3. What are the three basic pipe systems? ____

   ____

   ____

   ____

4. What is a heating zone? ____________________

____

Circle *T* for true or *F* for false.

5. T F Circuit breaker boxes are usually located in each room of the house.
6. T F Residential structures hook their plumbing into a sewer system or a septic system but never both.
7. T F A condenser cools the air for air-conditioning systems.

CHAPTER 11

1. What are moldings? ____________________

____

2. Name three bathroom fixtures. ____________________

____

3. What is a sump pump? ____________________

____

4. Where are shut-off valves put? ____________________

____

5. What is track lighting? ____________________

____

Circle *T* for true or *F* for false.

6. T F Every room in the house usually has a chandelier.
7. T F Bathrooms usually have at least some waterproof surfaces.
8. T F Wood is stained to darken and protect it.

CHAPTER 12

1. Name three steps in laying down a brick walkway. ____________________

____

2. What is blacktop? ____________________

____

3. What is sod? ____________________

Circle *T* for true or *F* for false.

4. T F Landscapers always recommend fast-growing shrubs.
5. T F Lawns can contain many different types of grass.
6. T F Fencing is usually only used for decoration.

CHAPTER 13

1. When large equipment moves, what usually warns workers of the movement? ______________________________

2. Why is the type of clothing worn a safety factor? ______________

____

3. What are two safety factors involved in using ladders and scaffolds?

____

____

PART 1 Answer Key

SELF-CHECK 1-1

Answers will vary.

SELF-CHECK 1-2

1. Answers will vary.
2. Answers will vary.
3. Answers will vary.

SELF-CHECK 1-3

Answers will vary, but students should include the index, glossary, or list of figures.

SELF-CHECK 1-4

1. Reading to do
2. Reading to learn
3. Reading to learn
4. Reading to do
5. Reading to do
6. Reading to learn
7. Reading to learn
8. Reading to do
9. Reading to learn
10. Reading to learn

SELF-CHECK 1-5

1. Answers will vary.
2. Answers will vary.
3. Answers will vary.
4. Answers will vary.
5. Answers will vary.

SELF-CHECK 1-6

Answers will vary.

SELF-CHECK 1-7

1. Answers will vary.

2. Answers will vary.
3. Answers will vary.
4. Answers will vary.
5. Answers will vary.

SELF-CHECK 1-8

1. Answers will vary.
2. Answers will vary.
3. Answers will vary.
4. Answers will vary.
5. Answers will vary.

PART 2 # Answer Key

SELF-CHECK 2-1

1. 13
2. Framing the Building
3. 4
4. Framing the Floor
        Subflooring
        Finishing the Floor
    Framing the Walls
        Outside Walls
        Inside Walls
    Roughing In the Mechanicals
        Plumbing
        Electrical Work
    Framing the Ceilings and Roof
        Ceilings
        Roof Framing
5. page 15
6. Wood Exteriors: Chapter 7

    Stairs: Chapter 9
7. Answers will vary, but students should say that it lists the topics covered in the book and gives the page numbers on which the discussion of each topic begins.

SELF-CHECK 2-2

1. 8
2. Figure 3-6
3. page 53
4. Chapter 11
5. Chapter 2
6. page 120
7. Answers will vary, but students should say that a list of figures helps them locate the chapter where the figure appears and the figure's placement within the chapter.

SELF-CHECK 2-3

1. a. Married; single

b. reg. rate and gross pay

2. a. 3' 0"

   b. 36

   c. ¾"

SELF-CHECK 2-4

1. c
2. c
3. c

SELF-CHECK 2-5

1. 21, 59-76
2. 6
3. Walkways
4. bids and; permits and; planning by
5. 21, 38, 52, 55-57
6. 1

EXERCISES

THE TABLE OF CONTENTS

1. page 77
2. Chapter 7
3. 16
4. Materials of Construction
5. page 42

THE LIST OF FIGURES

1. 6
2. 8
3. page 30
4. A drawing detailing the second floor of a house.
5. page 79
6. Blowing in loose fill insulation.
7. Chapter 3
8. Figure 5-5
9. a. Symbols on a drawing.

   b. Several symbols used on a plan drawing.

c. Answers will vary.

10. a. Vinyl siding.

b. A home at a site with recently installed vinyl siding.

THE GLOSSARY

1. a. **form** A wooden or cardboard setup that holds poured concrete in shape until it dries.

   b. **lather** A worker skilled in putting a metal or wood structure on the wall of a building, which will hold plaster or stucco.

   c. **scale** A proportion, as on a drawing, of the relationship of one size to what it actually represents.

2. a. False

   b. False

   c. True

   d. False

THE INDEX

1. a. pages 85, 86-89

   b. pages 21, 38, 52, 55-57

   c. page 61

   d. page 72-73

   e. page 53-54

2. a. driveways

   tiles

   b. ceilings

   floor

   roof

   roughing in mechanicals and

   walls

PART 3 *Answer Key*

SELF-CHECK 3-1

1. Types of Construction
    - Residential Construction
    - Commercial Construction
    - Heavy Construction
    - Municipal Construction

   Jobs in Construction
    - The Building Trades
        - Carpenters
        - Ironworkers and Steelworkers
        - Bricklayers
        - Stonemasons
        - Cement Workers
        - Operating Engineers
        - Electricians
        - Plumbers and Pipefitters
        - Sheet-Metal Workers
        - Plasterers
        - Glaziers
        - Roofers
        - Painters and Paperhangers
        - Insulation Workers
    - Other Building Trades
    - Management
        - On-Site Managers
        - Off-Site Managers
    - Professionals
        - Architects
        - Draftspeople, or Drafters
        - Engineers
2. 2
3. 27
4. 4

SELF-CHECK 3-2

1. Answers will vary.
2. Answers will vary.
3. Answers will vary.

SELF-CHECK 3-3

Answers will vary. Terms should include *construction, residential, commercial, heavy, municipal, multi-use, bid, contractor,* and *contract.*

SELF-CHECK 3-4

1. Topic Sentence: Construction jobs fall into three general categories.

   Supporting Details: The largest, the building trades, employs all the unskilled and craft workers.

   Management employs both the managers on the site and those who manage the start-up, ordering, and financing of projects.

2. Topic Sentence: The workers in the building trades can be unskilled or highly skilled.

   Supporting Detail: Some workers learn their craft in technical schools or on the job.

SELF-CHECK 3-5

1. Possible Answer: Construction workers need to work safely. They must know how to use equipment and handle materials properly.
2. Possible Answer: Ironworkers and steelworkers frame large projects or they work on ornamental projects, such as fences, stairways, and gates.

SELF-CHECK 3-6

1. Possible Answers:

   There are four types of construction—residential, commercial, heavy, and municipal. The construction industry employs more workers than any other in the United States. Construction jobs fall into three basic categories—building trades, management, and professionals. The building trades require more knowledge than ever before. Construction work can be dangerous.

2. Answers will vary. Terms should include *site, construction, building codes, bids, contractor, craft, apprentice, carpenter, ironworkers, steelworkers, ornamental, bricklayers, stonemasons, mortar, cement, masons, concrete, operating engineers, electricians, plumbers, pipefitters, sheet-metal workers, plasterers, plaster, stucco, lather, glaziers, roofer, painters, paperhangers, insulation, general superintendents, job superintendents, foremen, forewomen, architect, draftsperson, renders, surveyor, drawings, plans,* and *engineers.*

3. Answers will vary.

PREVIEW

1. Working on a Construction Site
2. The General Contractor

   Job Superintendent

   Craft or Trade Workers

   Stages of Construction
3. Figure 2-1 Trailer office.

   Figure 2-2 Payroll register.

   Figure 2-3 Architect's plans.

   Figure 2-4 Inspector at site.

   Figure 2-5 Plumber's bid.

   Figure 2-6 General contractor's bid.

QUESTION

1. Answers will vary.
2. What is a general contractor?

   What is a job superintendent?

   What are craft or trade workers?

   What are the stages of construction?
3. Answers will vary.

READ

1. a. architect's

   b. drawings

   c. zoning officials

   d. family

   e. drawings
2. Answers will vary. Terms should include *general contractor, bid, subcontractors, plumbers, electricians, architect's plans, drawings,* and *zoning officials.*
3. Topic Sentence: The skilled workers at a construction site fall into one of three categories.

   Supporting Details: Structural workers lay the foundation and build the frame. They also do the masonry and ironwork. Mechanical workers do the plumbing, heating, air conditioning, elevators, and sheet-metal work. Finishing workers finish the inside walls and floors. They also finish the insulation, roofing, windows, and siding.

4. Possible Answer: There are three types of skilled construction workers at the site. The foundation, frame, masonry, and ironwork is done by structural workers. Mechanical workers do the work having to do with mechanical devices. The finishing work, such as inside walls, insulation, roofing, and siding, is done by finishing workers.
5. Topic Sentence: Every construction project starts with planning.

   Supporting Details: The first step is to plan the type of project and location. The next step involves the design of the project. Usually, an architect supplies plans. The plans must meet government, usually local, approval.

RECODE

1. Possible Answers:
   a. The job of general contractor is to get and submit bids. Once the bid is awarded, the planning starts.
   b. Jordan hires all on-site workers and subcontractors. Subcontractors do just one specialized part of the job, such as plumbing.
   c. Jordan follows the architect's plans and makes sure everyone else does too. The plans are a set of drawings.
2. There are always jobs for skilled craft workers in construction. Often, local government standards and codes say that these workers must be the only ones to perform certain tasks.

REVIEW

1. Answers will vary.

CHAPTER 3

PREVIEW

1. Designing the Project
2. The Architect's Role

   Reading Plans
3. Putting the Design on Paper

   Getting Bids

   Getting Permits
4. Figure 3-1 A multi-use project.

   Figure 3-2 A preliminary design.

   Figure 3-3 A detail drawing.

   Figure 3-4 Electrical plans.

   Figure 3-5 A specifications page.

   Figure 3-6 A scale on a drawing.

QUESTION

1. What is involved in designing the project?
2. Answers will vary.
3. What is the architect's role?
4. Answers will vary.
5. What is involved in reading plans?
6. Answers will vary.
7. A scale on a drawing. I would use the scale to figure out the actual sizes of the things pictured in the drawing.

READ

1. Answers will vary. Terms should include *drawing, plans, details, draftsperson, architect, owner,* and *building codes.*
2. **Paragraph 1**

   Topic Sentence: The owner of the project hires an architect.

   Supporting Details: The architect submits a design to the owner. Before the designs is started, certain things are taken into account. First, the owner defines the needs of the project.

   Paragraph 2

   Topic Sentence: If the project is a single-family home, the architect will consult with the owner.

   Supporting Details: Together, they determine what type of structure will be built. If the project is commercial, many other needs arise. Some projects are multi-use. This means they serve both residential and commercial uses. Figure 3-1 shows a completed multi-use project.

   Paragraph 3

   Topic Sentence: Next, the architect discusses budgets with the owner.

   Supporting Details: The architect determines the size, materials, location, and design based on the budget.

   Paragraph 4

   Topic Sentence: The architect then considers building codes and other municipal requirements.

   Supporting Details: If the location is in a historical district, only certain designs will be approved. Cities and towns have rules about how large a building may be. They also have many other restrictions. Some rules are designed to keep an area totally residential. Others consider the traffic patterns that some projects may generate. The architect must follow all of these requirements before the design is approved.
3. **Paragraph 1**

   Topic Sentence: After all the research, the architect is now ready to proceed.

   Supporting Details: The next step is a rough design. The architect shows the design to the owner. Figure 3-2 shows a preliminary design.

Paragraph 2

Topic Sentence: Usually, the owner makes changes and suggestions.

Supporting Details: Then the architect will make further roughs of the design. The architect and the owner reach a basic agreement about the design.

Paragraph 3

Topic Sentence: The architectural firm will then produce a set of plans.

Supporting Details: Usually, a draftsperson does the actual drawings. In technical schools, most students take a basic drafting course. Some students choose to specialize in drafting. Others go on to specialized building trades. Being familiar with drafting is a basic construction necessity.

Paragraph 4

Topic Sentence: The set of plans details each floor and room.

Supporting Details: It shows the doors, windows, closets, ceilings, electrical outlets, plumbing, and any other details. Figure 3-3 shows a drawing of the detail of the second floor of a house.

4. **Paragraph 1**

Topic Sentence: Sometimes, the architect will help during the pricing stage.
Supporting Details: If the builder is a general contracting firm, then the firm supplies the bids. If the owner is an individual, the architect will often review all bids. Every phase of the project has a cost. Every item used also has a cost.

Paragraph 2

Topic Sentence: All contractors and suppliers base their prices on the plans.

Supporting Details: Often, several prices, or bids, are gotten so that costs can be compared.

Paragraph 3

Topic Sentence: Builders do not always take the lowest bid.

Supporting Details: They also consider reputation for quality. In addition, contractors must have the ability to complete the project on schedule and within the bid. Sometimes, the lowest bid can cost more in the end because the contractor makes many mistakes.

5. Topic Sentence: The architect and builder both work on the next stage.

Supporting Details: Before any building starts, the project needs a building permit. A building permit is issued by an authorized government agency. Usually, the zoning commission has reviewed plans. They may have made some suggestions as to locale, historical features, and so on. The final plans are submitted. If the zoning commission approves, a building permit is written up. It must be posted at every site.

RECODE

1. Answers will vary.

2. Possible Answers:

   a. If a living-room plan calls for a fireplace, the specifications will give the materials and their exact size. In the fireplace discussed here, the marble is 1/2 an inch thick and 16 inches wide.

   b. The drawings of the fireplace tell the contractor what materials to include in the bid. They also show the carpenter what space to allow. The mason does the actual building of the fireplace.

   c. The drawings are to scale. The scale is a ratio. It shows a much large item in a small drawing.

REVIEW

1. Answers will vary.
2. The defined boldface terms are as follows.

   building permit A permit, usually issued by a local governmental department, to proceed with a construction project.

   detail drawing A drawing in detail of a part of a structure or project.

   drafting The drawing of the specifications of structures.

   drawing An illustration, particularly a page in a set of plans.

   elevation A scale drawing of one view of a structure or part of a structure, as of the front, side, or rear.

   plot plans showing the site for a building and the specifications for excavation.

   scale A proportion, as on a drawing, of the relationship of one size to what it represents.

   section view A drawing that shows the internal detail of a structure or a particular part of a structure.

   specifications A detailed statement of something, such as a structure, an item, or how some work is performed.

CHAPTER 4

PREVIEW

1. Construction Tools and Materials
2. Materials of Construction

   Construction Tools
3. Exterior Materials

   Interior Materials

   Fasteners and Adhesives

   Interior Furnishings

4. Figure 4-1 Weathered wood.

   Figure 4-2 Vinyl siding.

   Figure 4-3 Fancy bathroom.

   Figure 4-4 Installing wallboard.

   Figure 4-5 Several measuring devices.

   Figure 4-6 Several planes.

   Figure 4-7 Using a hammer.

   Figure 4-8 Portable circular saw.

   Figure 4-9 Portable electric drill.

QUESTION

1. What are construction tools and materials?
2. Answers will vary.
3. What kinds of materials are used in construction?
4. Answers will vary.
5. What kinds of tools are used in construction?
6. Answers will vary.
7. A portable circular saw. A carpenter uses it to cut wood and other materials at the site.

READ

1. Answers will vary. Terms should include *design, tools, materials, concrete, steel, foundations, footings, framing, wood, steel-reinforced concrete slabs,* and *climate.*
2. **Paragraph 1**

   Topic Sentence: Once the basic framing is done, a variety of other materials will be needed.

   Supporting Details: Exterior finishes vary greatly. In certain climates, insulated boards are attached to the framing. Some sort of siding is put in on top of them. In other climates, a thinner board is attached.

   Paragraph 2

   Topic Sentence: Siding also varies by climate and location.

   Supporting Details: Near seawater, certain kinds of wood weather well. Figure 4-1 shows weathered wood on a seaside cottage. In sunny climates, stucco handles the extremes of heat well.
3. **Paragraph 1**

   Topic Sentence: Once inside the structure, one finds an even greater variety of materials.

Supporting Details: In commercial buildings, walls, ceilings, and floors take a lot of wear. Most commercial materials meet certain standards for wear. For instance, commercial carpeting can withstand thousands of people crossing over it daily. Such carpeting is often inappropriate for a home in which only a few people live.

Paragraph 2

Topic Sentence: The use of interior materials depends greatly on price.

Supporting Details: A bathroom can be a simple white tile room with minimum space and the basic necessities. Or, it can include a whirlpool, steam room, a spacious area, and many custom-built features. The difference in price may be enormous. Figure 4-3 shows a lavish bathroom.

Paragraph 3

Topic Sentence: Taste and use are other factors in choosing materials.

Supporting Details: A family that enjoys cooking at home a lot may devote much money to a gourmet kitchen. Someone who does not cook may need only the most practical kitchen.

Paragraph 4

Topic Sentence: Interior walls may also vary greatly.

Supporting Details: Again, cost, taste, and use determine the choice of materials.

RECODE

1. Possible Answers:

   a. Fasteners are devices, such as nails, screws, bolts, and staples, that hold materials together. Different fasteners are used depending on the materials and their placement.

   b. Adhesives bond two items. They are also used to strengthen bonds made with fasteners.

2. Possible Answers:

   Architectural plans do not usually include furniture. They do usually include everything that is built in, such as shelves and appliances. Fixtures are usually included on the plans.

3. Possible Answer: There are four main types of hand tools—measuring or marking, cutting, drilling or boring, and fastening.

4. Possible Answers:

   a. Jim tells you that you have seen the hand tools. However, a large construction job usually requires lots of power tools.

   b. Power tools can be portable or held and operated by hand. They can also be stationary, or set in one place. There are three types of power tools—cutting, drilling or boring, and fastening.

REVIEW

1. Answers will vary.

CHAPTER 5

PREVIEW

1. Preparing the Site and Starting
2. Finding the Best Location

   Preparing the Site

   Putting in the Foundation
3. Footings

   Slabs

   Foundation Walls
4. Figure 5-1 Leveling the site.

   Figure 5-2 A surveyor at work.

   Figure 5-3 A bulldozer digging a foundation.

   Figure 5-4 The work area around a foundation.

   Figure 5-5 Forms for holding concrete.

QUESTION

1. What happens in preparing the site and starting?
2. Answers will vary.
3. How do you find the best location?
4. Answers will vary.
5. How do you prepare the site?
6. Answers will vary.
7. How do you put in the foundation?
8. Answers will vary.
9. Heavy equipment leveling a site. Answers will vary.

READ

1. Answers will vary. Terms should include *site, leveling, heavy equipment,* and *foundation.*
2. **Paragraph 4**

   Topic Sentence: Also, certain soils contain chemicals that decay some foundation materials.

Supporting Details: A special concrete mixture can protect the foundation from decay.

Paragraph 5

Topic Sentence: The actual location of the structure depends on many factors.

Supporting Detail: The contractor needs access to the site. Building codes determine placement of sewer lines or septic fields. The soil condition also has a role.

Paragraph 6

Topic Sentence: The architect will have determined the location.

Supporting Details: The contractor may run into problems. Some problems require changing the exact location.

3. **Paragraph 1**

Topic Sentence: When Bardel and Baker contracted to build the four homes, surveyors came in and marked the four lots.

Supporting Details: The marks show the exact outside lines of the lots. The architects inspected the lots. They chose the site of each home.

Paragraph 2

Topic Sentence: Surveyors use exact instruments.

Supporting Details: They must have certain qualifications. A survey becomes a legally acceptable document once it is completed. Figure 5-2 shows a surveyor at work.

Paragraph 3

Topic Sentence: To find the exact location of each home, architects and engineers look at many factors.

Supporting Details: Does the house need a septic system? Higher ground may help avoid basement moisture. One location may save more trees than another. Access to the road may be easier in one placement. How the home faces the sun and how it faces the neighbors is also important.

4. Topic Sentence: Lily tells you, "Our next step is to put down footings.

Supporting Details: Footings are small concrete supports poured below the level of the foundation. The footings hold the floor support posts. The plans show the number of footings. They also show where the footings are to be placed."

RECODE

1. Possible Answers:

a. The type of foundation depends on the soil and the climate. A slab always forms the base of the house. In warm climates, a basement is not needed.

b. The concrete for the slab is poured. Other materials can be used. Some slabs take several pours. Some are extra thick.

c. The mechanical lines have to be put in before the slab is poured. The slab is poured after forms are built to hold the shape of the slab. After the slab dries, the forms are removed.

   d. Forms can be wood or other materials, such as cardboard. Some forms are prefabricated.

2. Possible Answers:

   a. The foundation walls are often made up of concrete blocks. There are support posts built into the foundation. The concrete blocks are held together with mortar and are placed in alternating patterns for strength.

   b. Some foundation walls are made from poured concrete. The concrete is poured into large forms which are left in place until the concrete dries.

REVIEW

1. Answers will vary.
2. The defined boldface terms are as follows:

   excavation The removal of soil by digging as for a foundation.

   footing A supporting base, such as a wall, usually made of concrete poured into a form.

   form A wooden or cardboard setup that holds poured concrete in shape until it dries.

   foundation The base of a building that supports the structure.

   level To even out.

   prefabricated Constructed in advance, as in a part of a house that is brought to a site in one piece.

   rough in To allow space and do preliminary work, as for the mechanicals of a structure.

   septic system A sewage disposal system in which wastewater flows into a tank where it is decomposed.

   site The location of a construction project.

   slab A fairly thin, large piece of material, such as concrete.

   surveyor A worker skilled in the exact measuring of land.

REVIEW YOUR KNOWLEDGE

CHAPTER 1

1. Construction of structures for housing.
2. A beginner who assists an experienced person in return for learning a trade.
3. Carpentry work.
4. Plumbers and electricians.
5. b.
6. b.
7. F
8. F

CHAPTER 2

1. A builder who contracts for an entire project.
2. A worker who works on the frame, foundation, and structural iron and masonry.
3. A worker who puts in the lines for mechanical devices, such as electricity, plumbing, and heating.
4. Planning.
5. The building of the underlying structure of a project.
6. a.
7. d.
8. T
9. T

CHAPTER 3

1. One who designs and supervises the construction of projects, such as homes, buildings, bridges, and so on.
2. A person skilled in drafting.
3. A permit, usually issued by a local government agency, to proceed with a construction project.
4. A drawing that shows the internal detail of a structure or a particular part of a structure.
5. A proportion, as on a drawing, of the relationship of one size to what it actually represents.
6. d.
7. a.
8. F
9. T

CHAPTER 4

1. Wood.
2. Steel and concrete.
3. Wood and vinyl.
4. Any of several types of sheets or boards used to construct walls.
5. Nails, screws, bolts, and staples.
6. d.
7. d.
8. F
9. T

CHAPTER 5

1. The location of a construction project.
2. The foundation.
3. Make exact measurements of land and maps.
4. A wooden or cardboard setup into which concrete is poured.
5. b.
6. c.
7. F
8. T

PART 4 *Answer Key*

SELF-CHECK 4-1

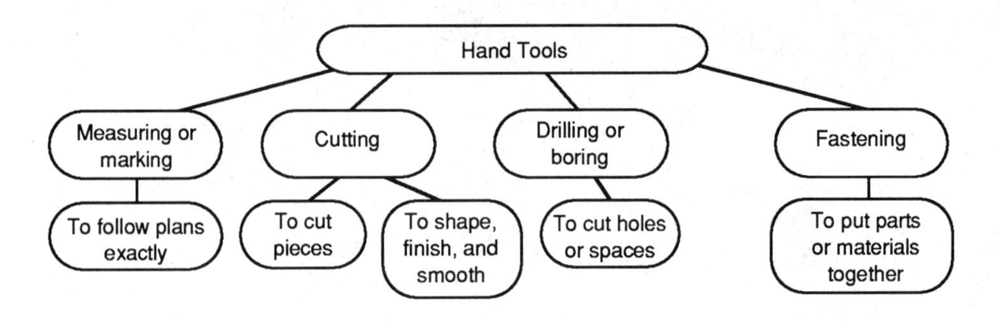

SELF-CHECK 4-2

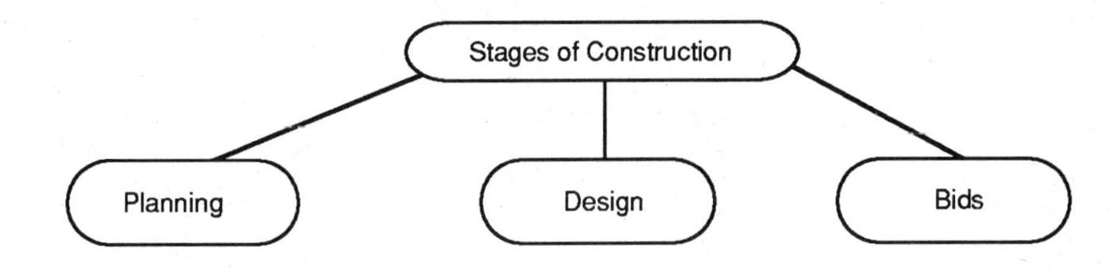

SELF-CHECK 4-3

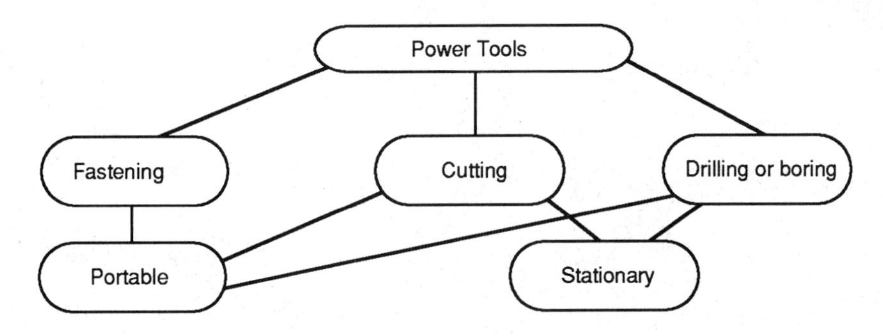

SELF-CHECK 4-4

B. Management
   1. On-site
   2. Off-site
C. Professionals
   1. Architects and engineers
   2. Other technicians

II. (Next Main Idea)

SELF-CHECK 4-5

I. Job Superintendent

   A. Large projects

      1. Reports to general superintendent

      2. Oversees only small portion

   B. Small projects

      1. In charge

      2. Does actual work

   C. Hires and fires

      1. Unskilled workers

      2. Skilled workers report to subcontractor

II. (Next Main Idea)

PRACTICING PQ3R WITH CHAPTER 6

PREVIEW

1. Framing the Building
2. Framing the Floor
   Framing the Walls
   Roughing In the Mechanicals
   Framing the Ceilings and Roof
3. Subflooring
   Finishing the Floor
4. Figure 6-1 The frame of a floor.
   Figure 6-2 A floor truss system.
   Figure 6-3 A subfloor.
   Figure 6-4 A plan for framing an exterior wall.
   Figure 6-5 Interior walls being framed.
   Figure 6-6 Symbols on a drawing.
   Figure 6-7 The six basic roof styles.
   Figure 6-8 The frame of a roof.

QUESTION

1. a. Answers will vary.
   b. Answers will vary.
   c. Answers will vary.
   d. Answers will vary.

READ

1. (a) anchor bolts, (b) plates, (c) building, (d) shield.

2. Possible Answer: Workers put anchor bolts in the foundation so that the sill plate can be held. A termite shield is required by code.

3. **Paragraph 1**

   Topic Sentence: The split-level requires a two-level subfloor.

   Paragraph 2

   Topic Sentence: The plywood base is called a subfloor.

4. a. F

   b. F

   c. T

5. **Paragraph 1**

   Possible Answer: Carpenters frame walls in a number of different ways. They follow the plans for window and door openings.

   Paragraph 2

   Possible Answer: Studs are shown on the plan. The studs are upright boards that form the frame of the walls.

   Paragraph 3

   Possible Answer: The exterior walls have to be strong enough to protect the house from the weather and to hold the insulation and interior finishing materials.

6. (a) partitions, (b) studs, (c) roughed in, (d) openings.

7. **Paragraph 1**

   Possible Answer: The plumber puts all the pipes into the walls and floors during framing.

   Paragraph 2

   Possible Answer: During the roughing in, pipes and ducts are installed but not yet connected.

   Paragraph 3

   Possible Answer: The plumber can get the work done easily before the walls and floors are finished. All the rough work is covered later during finishing.

8. (a) roughs in, (b) electricity, (c) inspectors

9. a. T

   b.F

10. (a) six, (b) gable, (c) hip, (d) gambrel

RECODE

1.

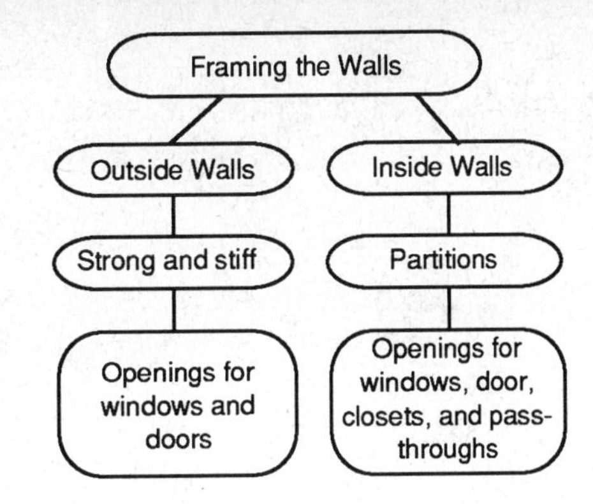

2.

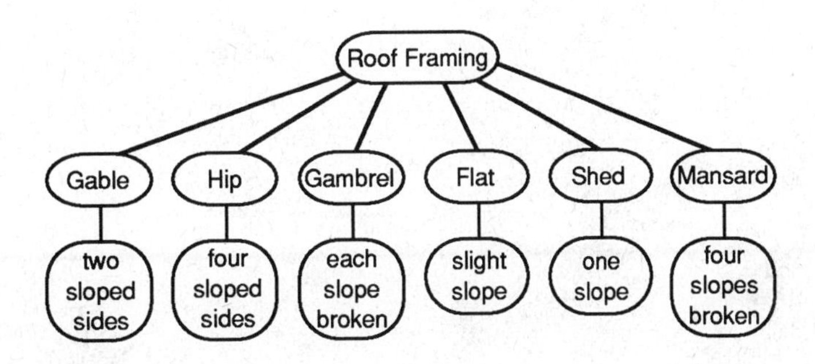

REVIEW

1. The defined boldface words are as follows.

anchor bolt A large bolt used to fasten a sill plate to a foundation or footing.

beam A length of lumber or metal, usually supported at each end, used in construction to support a load.

dormer A projection in a sloping roof that holds a window.

flat roof A roof having virtually no slope.

gable roof A pitched roof that has one or more gables. Gables are triangular wall sections that have a sloped top and stick out of a pitched roof.

gambrel roof A roof that has two slopes on each side. The lower slope has the deeper pitch.

girder A horizontal metal or wood beam used as a main support.

hip roof A roof having sloping edges and sides.

joist The horizontal timber to which the floor or ceiling board is nailed.

lean-to roof A single-pitched roof, usually attached to a shed.

mansard roof A roof that has two slopes on all four sides.

partition A divider, especially an interior wall.

post A piece of wood or other material set upright and used as a support.

rafter One of the sloped beams of a roof.

rough in To allow space and do preliminary work for, as for the mechanicals of a structure.

shed roof A roof with a slope in one direction only.

stud An upright post in the framework of a building.

subfloor Rough boards laid over the joists on which the finished flooring is laid.

truss system A prefabricated system of trusses meant to go together easily at the site.

PRACTICING PQ3R WITH CHAPTER 7

PREVIEW

1. Enclosing the Exterior
2. Masonry Exteriors

   Wood Exteriors

   Aluminum or Vinyl Siding

   Cement and Stucco Exteriors
3. Figure 7-1 Applying a brick veneer.

   Figure 7-2 Horizontal wood siding.

   Figure 7-3 Vertical wood siding.

   Figure 7-4 Wood shakes.

   Figure 7-5 Installing vinyl siding.

   Figure 7-6 Applying a stucco exterior.

QUESTION

1. a. Answers will vary.

   b. Answers will vary.

   c. Answers will vary.

   d. Answers will vary.

READ

1. **Paragraph 1**

   Possible Answer: Bricklayer Hiroko Hanaka is building a brick wall and will build a brick fireplace on the rear wall.

Paragraph 2

Possible Answer: Hiroko is really using a brick veneer, which is surface brick attached to the frame wall.

Paragraph 3

Possible Answer: Hiroko spreads mortar between the bricks to hold them in the alternating pattern.

2. **Paragraph 1**

   Topic Sentence: Wood siding remains one of the most common siding materials.

   Supporting Details: Wood siding can be applied horizontally or vertically.

   Paragraph 2

   Topic Sentence: Horizontal wood siding is usually manufactured to specific widths and lengths.

   Supporting Details: These pieces are nailed to the sheathing on the outside of the house frame. Figure 7-2 shows horizontal wood siding.

3. (a)vinyl siding, (b)wood, (c) cement, (d) stucco, (e)mesh.

RECODE

1. Possible Answer: Masons or bricklayers can work with stone and brick. Most brick is now veneer. Walls made entirely of brick are unusual. The bricklayer spreads mortar between bricks that have been set in an alternating pattern.

2. **Paragraph 3**

   Possible Answer: The installation of vertical siding is similar to that of wood siding. Vertical siding can be used to create different designs.

   Paragraph 4

   Possible Answer: Shingles and shakes come already nailed to a panel or can be nailed in different patterns. Prefabricated panels are easy to use but cost more.

   Paragraph 5

   Possible Answer: Wood can be stained, painted, or left to weather. The effect of weathering depends on the climate.

3.

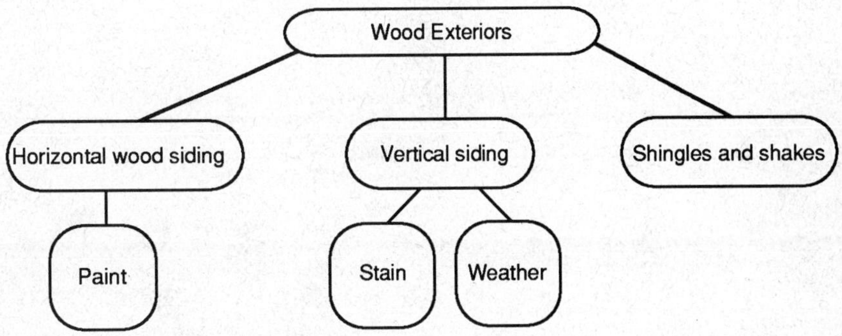

Wood Exteriors
Horizontal wood siding
Vertical siding
Shingles and shakes
Paint
Stain
Weather

REVIEW

1. The defined boldface words are as follows.

   mortar An adhesive used in joining items in masonry.

   shake A shingle with a rough surface, used for barns or to give a rustic look.

   shingle A thin oblong piece of material, such as wood or asbestos, laid in overlapping rows to cover roofs or sides of houses.

   veneer A thin piece of finishing material that is applied to the surface of a less expensive material.

   wood siding A wood material used for the exposed surface of walls and buildings, which can be stained, painted, or left to weather.

PRACTICING PQ3R WITH CHAPTER 8

PREVIEW

1. Insulation

   Walls

   Ceilings

2. 7

3. Figure 8-1 Installing blanket insulation.

   Figure 8-2 Blowing in loose fill insulation.

   Figure 8-3 Putting up a vapor barrier.

   Figure 8-4 Nailing drywall with a power nailer.

   Figure 8-5 Taping drywall.

   Figure 8-6 Ceramic kitchen tile.

   Figure 8-7 A suspended ceiling.

QUESTION

1. a. Answers will vary.

   b. Answers will vary.

   c. Answers will vary.

READ

1. **Paragraph 3**

   Topic Sentence: Ben is using blanket insulation.

   Paragraph 4

   Topic Sentence: Another type of insulation, loose fill, consists of loose material.

2. (a) foil, (b) heat, (c) vapor barrier, (d) condensation.

3. **Paragraph 1**

   Topic Sentence: Almost all the wall products can also be used to cover ceilings.

   Supporting Details: However, some of the ceramic tiles may be unsuitably heavy.

   Paragraph 2

   Topic Sentence: One additional material, ceiling tile, enjoys great popularity.

   Supporting Details: Ceiling tile can be suspended from the ceiling frame to lower a ceiling. Some ceiling tile has good acoustical properties, meaning that it helps to lower noise. Figure 8-7 shows a suspended ceiling.

   Paragraph 3

   Topic Sentence: Some ceilings are open to the roof framing.

   Supporting Details: This exposes the beams. Some people like the added height and the natural look of the wood.

RECODE

1.

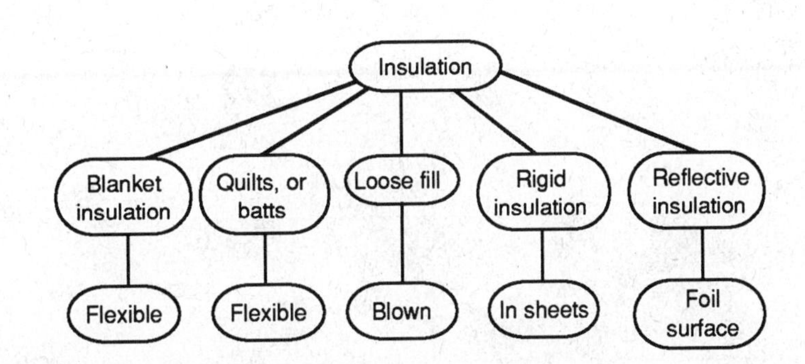

2. Possible Answer: The drywall covers all the framing and roughed in mechanicals. The walls also help keep sound in or out of a room Some designs call for additional soundproofing. The drywall is nailed against the studs. Then the joints are filled in and taped. A number of different finishes can be used for walls. Ceramic tiles, plaster, and laminates are just a few.

3. I. Ceilings
   - A. Ceiling tile
     - 1. Very popular
     - 2. Good acoustical properties
   - B. Some ceilings open to roof framing

   II. (Next Main Idea)

REVIEW

1. The defined boldface words are as follows.

   blanket insulation that comes in blanketlike strips.

laminated Covered with a thin sheet of finished material.

loose fill A type of insulation that consists of loose particles blown into a space.

plaster A mixture of lime, sand, water, and other materials that hardens to a smooth surface. Used for walls and ceilings.

quilt Insulation that comes in square pieces called batting.

reflective insulation with a surface that reflects heat into or out of a space.

rigid insulation that comes in the form of unbendable sections.

R-value A measure of a substance's ability to slow down or block the flow of heat.

vapor barrier A material or paint that prevents the passage of moisture into walls.

PRACTICING PQ3R WITH CHAPTER 9

PREVIEW

1. Adding Floors and Stairs
2. Stairs

   Floors
3. Figure 9-1 Stringers installed for stairs.

   Figure 9-2 Stairs with a modern newel post and balusters.

   Figure 9-3 Spiral stairs.

   Figure 9-4 Disappearing stairs.

   Figure 9-5 A tongue-and-groove floor installation.

   Figure 9-6 Installing hardwood tiles.

   Figure 9-7 Linoleum installation.

   Figure 9-8 Ceramic tile installation.

QUESTION

1. a. Answers will vary.

   b. Answers will vary.

READ

1. Possible Answer: Stairs are a very important design element. They come in a wide range of styles. They need to be built to exact dimensions for safety. There are rules for the spacing of risers and treads.
2. (a) weatherproof, (b) Concrete, (c) access, (d) fold up
3. Topic Sentence: Flooring comes in a wide variety of materials.

Supporting Details: Some rooms, such as bathrooms, customarily have water-resistant tile floors. Other rooms may have carpeting. Kitchens often have linoleum or tile floors. However, the most common flooring material is still wooden planks. Wood tiles have become popular in recent years.

RECODE

1. Possible Answer: Most floor installers follow exact installation rules. The installer has to be experienced to do a good job.

2.

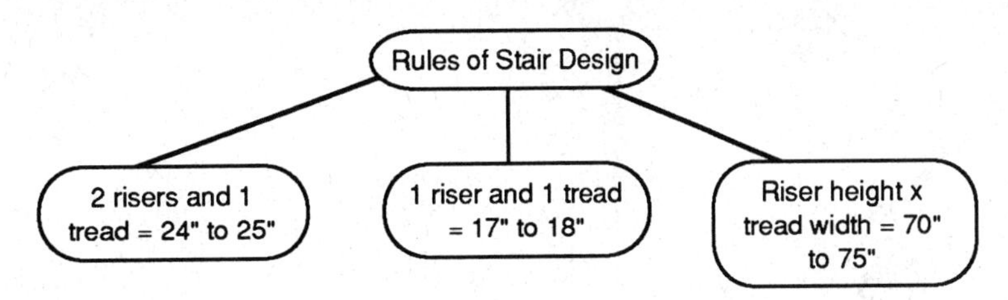

3. I. Floors
    - A. Water-resistant
        1. Linoleum
        2. Ceramic tile
    - B. Carpeting
    - C. Wood floors
        1. Wood planks
            - a. Various widths and thicknesses
            - b. Tongue-and-groove
            - c. Parquet
        2. Wood tiles
        3. Sanding
        4. Finishing

   II. (Next Main Idea)

REVIEW

1. The defined boldface words are as follows.

   baluster A vertical member of a stair railing used between the top and bottom rails.

   flooring Any of a number of materials used to construct floors.

   hardwood The wood of certain flowering trees, generally considered to have the greatest hardness.

   newel An upright post at the top and bottom of stairs.

   parquet floors Patterned wood floors, usually forming geometrical designs.

planking A group of heavy, thick boards.

riser The vertical part of a single step in a set of stairs.

stringer The sloping member that supports stairs.

tread The horizontal part of a stair step.

2. Answers will vary.

REVIEW YOUR KNOWLEDGE

CHAPTER 6

1. anchor bolts
2. A piece of wood or other material set upright and used as a support.
3. A horizontal metal or wood beam used as a main support.
4. The subfloor is usually made of plywood and is meant to hold the finished floor which is tile, wood, linoleum, or carpeting.
5. d.
6. T
7. F
8. F
9. F
10. T

CHAPTER 7

1. A thin piece of finishing material that is applied to the surface of a less expensive material.
2. Horizontal wood siding, vertical wood siding, and shingles or shakes.
3. It is relatively inexpensive and comes in a variety of colors.
4. d.
5. T
6. F
7. F
8. F

CHAPTER 8

1. A measure of a substance's ability to slow down or block the flow of heat.
2. Insulation that comes in blanketlike strips.
3. Loose fill, rigid, and reflective insulation.

4. Sheets of material used to form walls. Applied in "dry" condition as opposed to plaster, which is applied "wet."
5. a. Sound transmission class

   b. A material or paint that prevents the passage of moisture into walls.
6. a.
7. T
8. F

CHAPTER 9

1. A vertical member of a stair railing used between the top and bottom rails.
2. The vertical part of a single step in a set of stairs.
3. The horizontal part of a stair step.
4. The rough boards laid over the joists on which the finished flooring is laid.
5. Flooring that is manufactured with alternating depressions (called grooves) and extensions (called tongues) that fit together.
6. Concrete and plywood
7. b.
8. T
9. F

PART 5 *Answer Key*

SELF-CHECK 5-1

1. Main Idea: Hand tools fall into four main categories.

   Supporting Details: Measuring or marking tools make it possible to follow plans exactly. Cutting tools allow the worker to cut pieces to the size needed. They also allow workers to shape, finish, and smooth materials. Drilling or boring tools make cut holes or spaces. Fastening tools aid in putting parts or materials together.

2. Pictures will vary.

SELF-CHECK 5-2

Roof Framing

Type of Roof	Major Characteristic
Gable	Pitched roof with triangular projections
Hip	Sloped edges and sides
Gambrel	Two slopes on each side
Flat	Having little or no slope
Shed	Slope in one direction only
Mansard	Two slopes on all four sides

SELF-CHECK 5-3

Types of Construction

Type	Portion of Total Jobs	Major Projects
Residential	Largest part	Housing
Commercial	Second largest part	Business structures
Municipal	Third largest part	Public works, such as sewers and sidewalks
Heavy	Smallest part	Large public projects, such as bridges, dams

SELF-CHECK 5-4

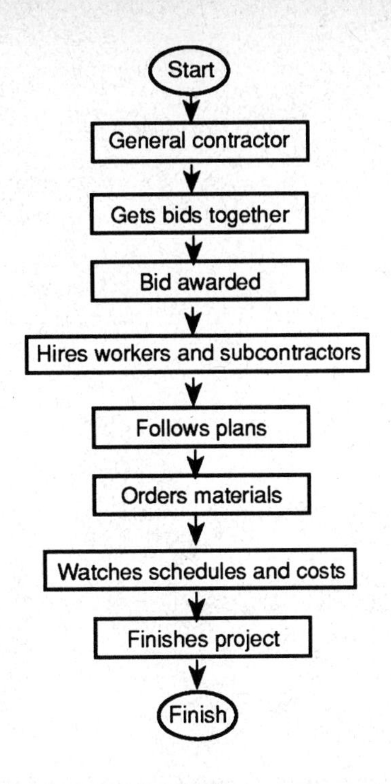

SELF-CHECK 5-5

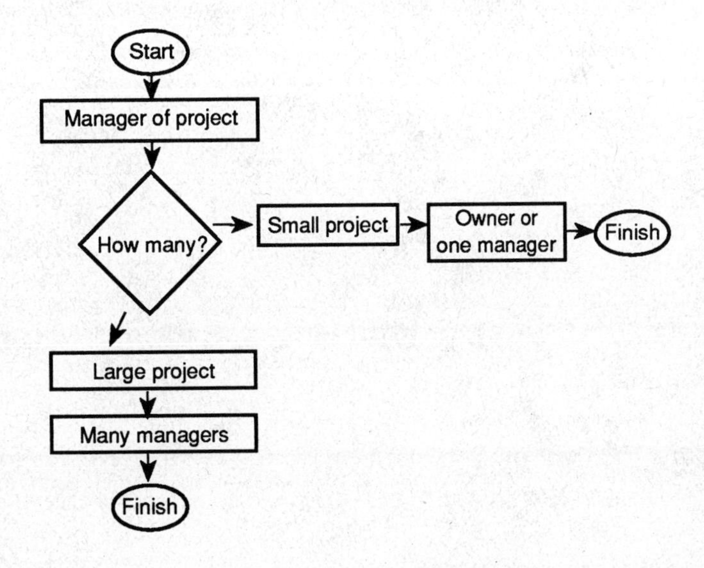

SELF-CHECK 5-6

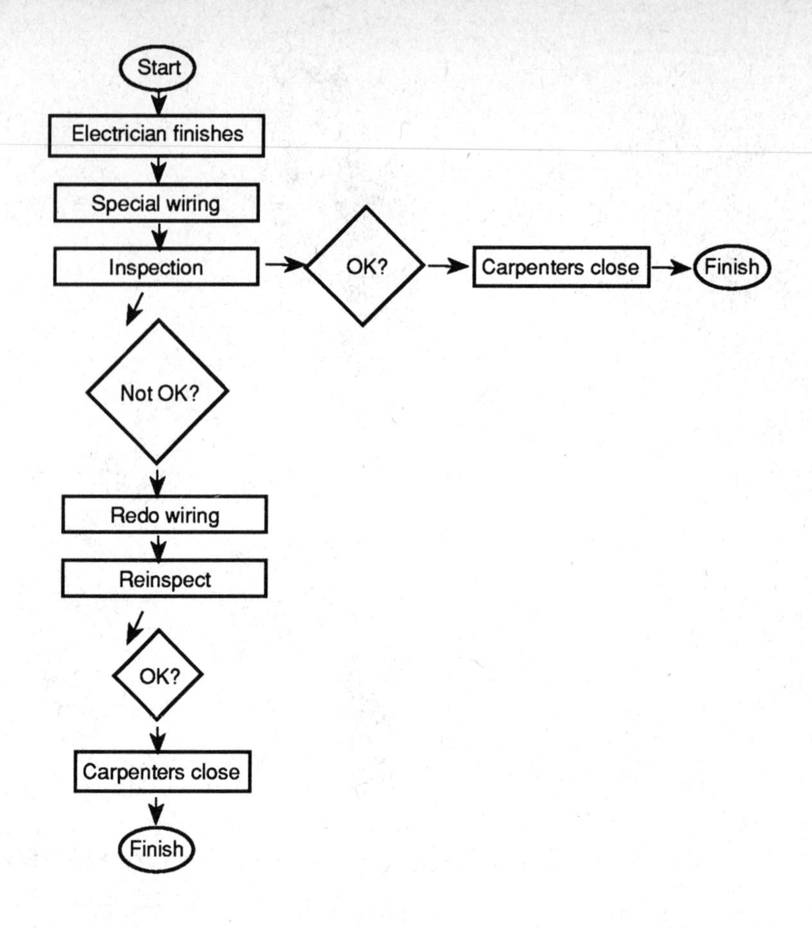

PRACTICING PQ3R WITH CHAPTER 10

PREVIEW

1. Plumbing, Electricity, and Heating
2. Electricity

   Plumbing

   Heating

   Air Conditioning
3. Figure 10-1 A circuit breaker box.

   Figure 10-2 Electrical connections being roughed in.

   Figure 10-3 Insulating pipes in an interior wall.

   Figure 10-4 A hot-water tank.

   Figure 10-5 A furnace.

   Figure 10-6 An air-conditioning condenser.

QUESTION

1. a. Answers will vary.

b. Answers will vary.

c. Answers will vary.

d. Answers will vary.

READ

1. **Paragraph 1**

   Topic Sentence: At the two-family home, electricians begin their work in the walls.

   Supporting Details: The exterior walls are up. The electricians have three days to install the wiring and outlets. After that, the carpenters come back in and put up the drywall.

   Paragraph 2

   Topic Sentence: Bogosi Electric got the subcontracting job from Baker and Bardel.

   Supporting Details: They often work with Baker and Bardel. Marsha and Jerry Bogosi own the company. One or the other is always at the job site supervising the electricians.

2. **Paragraph 1**

   Topic Sentence: The plumbers also arrive at the two-family.

   Paragraph 2

   Topic Sentence: "The pipes form a maze, just like the wires."

3. (a) pipes, (b) bathtubs, (c) showers, (d) hot-water tank, (e) wastewater

4. **Paragraph 4**

   Main Idea: Cold air is warmed in the furnace system which burns oil. The thermostats tell the system how much warm air is needed.

   Paragraph 5

   Main Idea: Registers installed in each room allow hot air to circulate.

RECODE

1. I. Cold-Water Pipes

   A. Bring fresh water

   1. Toilets

   2. Bathtubs

   3. Showers

   4. Sinks

   B. Separate line for ice maker

   II. Hot-Water Pipes

   A. Hot-water tank

   1. Electrically operated

   2. Heat from heating system

B. Bring hot water

1. Bathtubs
2. Showers
3. Sinks
4. Washing machines

III. Waste Pipes

A. Remove wastewater

B. Send wastewater out of house

1. Septic system
2. Sewer

IV. (Next Main Idea)

2. Possible Answer: The central air conditioning in all four houses requires a duct system. An outside condenser cools air when the inside thermostats send a signal. The duct system is put in during framing.

3.

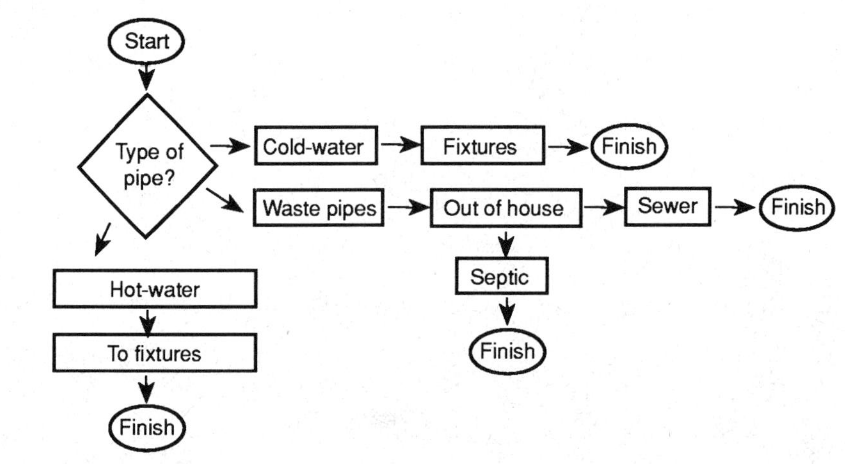

REVIEW

1. The defined boldface term is as follows.

plumbing The pipes, fixtures, and wastewater systems in a structure.

PRACTICING PQ3R WITH CHAPTER 11

PREVIEW

1. Interior Finishing
2. Adding Interior Trim

Fixtures and Appliances

Lighting

Finishing of Surfaces

Cabinets and Doors

QUESTION

1. a. Answers will vary.
   b. Answers will vary.
   c. Answers will vary.
   d. Answers will vary.
   e. Answers will vary.

READ

1. Interior moldings and trim make the house look finished.
2. **Paragraph 1**

   Topic Sentence: The architect specifies all fixtures and appliances.

   Paragraph 2

   Topic Sentence: The plumber also adds all valves and knobs to operate the fixtures.
3. (a) fixtures, (b) shut-off, (c) emergency, (d) water
4. **Paragraph 1**

   Main Idea: Lights are an important part of the design, and they come in a variety of styles.

   Paragraph 2

   Main Idea: The drawings show the location of all lighting fixtures, switches, and outlets.

   Paragraph 3

   Main Idea: Some fixtures can be put in before painting, but the fanciest fixtures are put up afterwards.
5. Topic Sentence: Door hardware can be complicated.

   Supporting Details: When the door closes, all hardware must meet exactly if the locks and knobs are to work. The carpenters make the hardware cutouts on the site so that they can match them as necessary. Figure 11-11 shows door hardware installation.

RECODE

1. Pictures will vary.
2. I. Lighting
   - A. Modern
     1. Recessed lighting
     2. Track lighting
   - B. Old-fashioned
     1. Wall sconces
     2. Chandeliers

II. Installation

A. Indicated on plans

B. Electricians install

III. (Next Main Idea)

3.

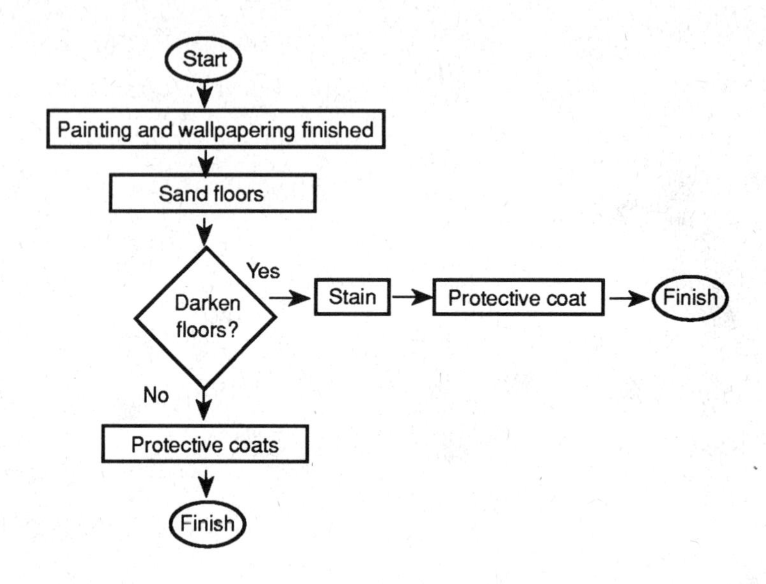

REVIEW

1. The defined boldface words are as follows.

   appliance A device meant for household use, usually operated by electricity or gas.

   fixture An item attached permanently for use in a building, such as sinks, toilets, lighting, and so on.

   molding An ornamental strip used to finish a surface, as the top of a wall.

   stain To color or darken with a coat of tint.

   sump pump A device that removes water from a depression, as in a basement floor.

   trim The moldings, hinges, locks, and any other ornamental parts that finish the areas around windows and doors, and other parts of a building.

2. Answers will vary.

PRACTICING PQ3R WITH CHAPTER 12

PREVIEW

1. Landscaping and Paving
2. Patios and Walkways

Driveways

Planting Trees and Shrubs

Lawns

Fencing

3. Figure 12-1 A brick walkway.

   Figure 12-2 A blacktop driveway.

   Figure 12-3 Shrubs and plantings alongside a house.

   Figure 12-4 Preparing ground for a new lawn.

   Figure 12-5 Fencing around a yard.

QUESTION

1. Answers will vary.
2. Answers will vary.
3. Answers will vary.
4. Answers will vary.
5. Answers will vary.

READ

1. Possible Answer: Juan Pato is a bricklayer. He is laying down a brick walkway in front of the house. First, he puts down sand. Then he puts the bricks in a pattern and wets them down.
2. Possible Answer: When the plans call for blacktop driveways, a subcontractor usually does the work. After all the trucks leave, blacktop is spread and must be left to dry for several days. A protective coating is put over the blacktop.
3. (a) plants, (b) lawn, (c) trees, (d)climate, (e) duration
4. Lawns can contain many different varieties of grass.

RECODE

1.

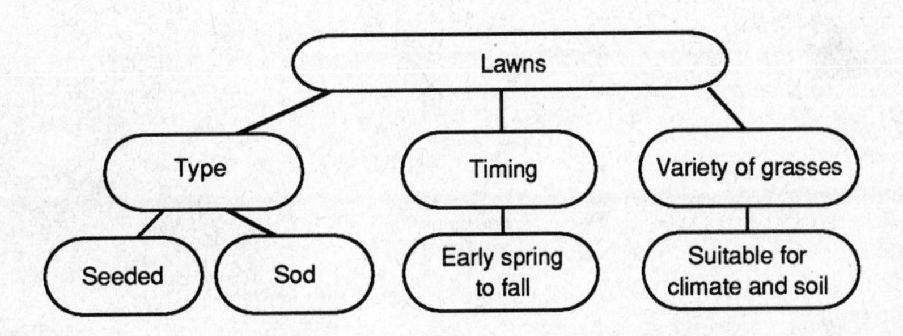

2.

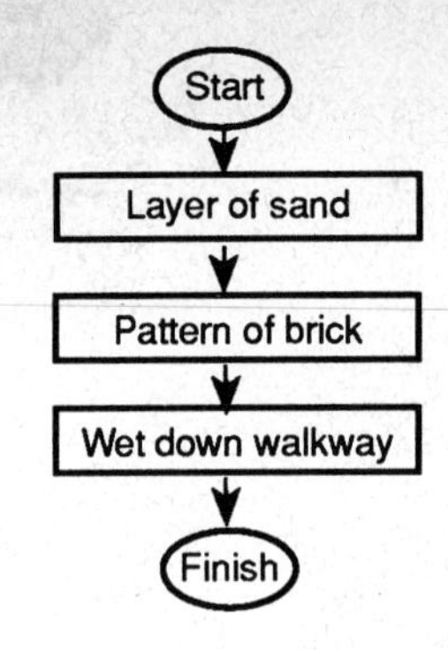

REVIEW

1. Answers will vary.

PREVIEW

PRACTICING PQ3R WITH CHAPTER 13

1. Safety on the Construction Site
2. Figure 13-1 Hard hats help prevent head injuries.

QUESTION

1. Answers will vary.

READ

1. Possible Answer: Workers in construction have to use ladders and scaffolds. Safe use is important. Ladders and scaffolds need to have strength and stability.

RECODE

1.

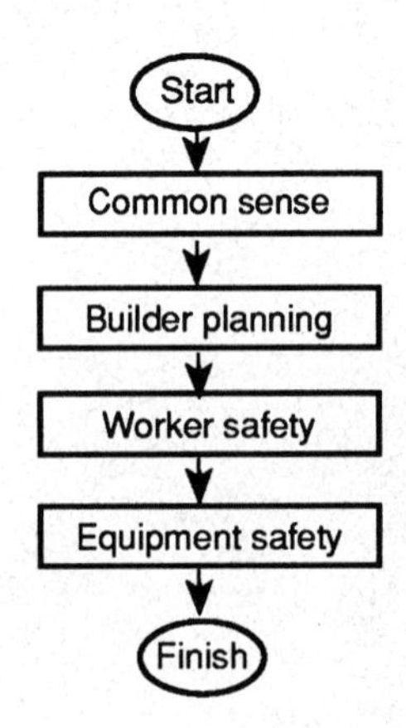

2. I. Construction Injuries

   A. Avoidable

      1. Use common sense

      2. Observe builder precautions

   II. Equipment Safety

      1. Always lock when not in use

      2. Use beepers to indicate reverse

   III. Working Injuries

      A. Lift properly

      B. Clean site

      C. Use equipment correctly

      D. Wear proper clothing and goggles

      E. Use ladders and scaffolds safely

   IV. (Next Main Idea)

REVIEW

1. Answers will vary.

REVIEW YOUR KNOWLEDGE

CHAPTER 10

1. During framing
2. The box in a house where the electricity comes into the house from the outside and is distributed through wires to various parts of the house.
3. Cold-water, hot-water, and wastewater
4. An area of the house controlled by one thermostat.
5. F
6. T
7. T

CHAPTER 11

1. Ornamental strips used to finish a surface, for instance, on the top of a wall.
2. Bathtubs, toilets, and sinks.
3. A device that removes water from a depression, as in a basement floor.
4. In every room that has a water supply.
5. Movable lights put on a track hung on the ceiling or wall.

6. F
7. T
8. T

CHAPTER 12

1. Putting down sand; laying bricks in a pattern; wetting down the walkway.
2. A substance that is spread over the ground to make a driveway.
3. Pre-grown lawn squares.
4. F
5. T
6. F

CHAPTER 13

1. A beeper.
2. Loose clothing can get caught in equipment.
3. Strength and stability.